DATE DUE			

10163

384.54
CAR

Careers in focus.
Broadcasting.

Broadcasting

Ferguson Publishing Company
Chicago, Illinois

Editorial Staff
Andrew Morkes, *Managing Editor-Career Publications*
Carol Yehling, *Senior Editor*
Anne Paterson, *Editor*
Nora Walsh, *Editor*

Copyright © 2003 Ferguson Publishing Company

Library of Congress Cataloging-in-Publication Data

Careers in focus. Broadcasting.-- 2nd ed.
    v. cm.
Includes index.
Contents: Art directors -- Broadcast engineers -- Camera operators -- Cartoonists and animators -- Disc jockeys -- Lighting technicians --Media planners and buyers -- Radio and television announcers -- Radio and television program directors -- Radio producers -- Real-time captioners -- Reporters -- Sports broadcasters and announcers -- Talent agents and scouts -- Television directors --Television editors -- Television producers -- Weather forecasters.
    ISBN 0-89434-440-4 (hardcover : alk. paper)
    I. Broadcasting--Vocational guidance--Juvenile literature. [I. Broadcasting--Vocational guidance. 2. Radio broadcasting--Vocational guidance. 3. Television--Vocational guidance. 4. Vocational guidance.] I. Title: Broadcasting.
HE8689.4 .C37 2003
384.54'023'73--dc21
                                        2002007656

Printed in the United States of America

Cover photo courtesy Bob Daemmrich/The Image Works

Published and distributed by
Ferguson Publishing Company
200 West Jackson Boulevard, 7th Floor
Chicago, Illinois 60606
800-306-9941
www.fergpubco.com

Table of Contents

Introduction

Broadcasting is the electronic transmission of images and/or sound. Radio and television are the main broadcasting media. The radio and television industry is made up of a large number of relatively small and independent stations that are individually owned and operated. Approximately 255,000 people are employed in both radio and TV. Commercial radio employs about half, and commercial television about one-third of the total. The others are employed by broadcasting headquarters, including network offices and public broadcasting stations.

Broadcasting is particularly exciting because it is like show business with a stopwatch; programming is timed down to the second, and precision and speed are crucial for both taped and live broadcasts. Broadcasting relies on the creativity of its employees to develop and hold the interest of its listening and viewing audiences. Yet, because of the unremitting pressures of deadlines, broadcasting must be geared to quick decisions and quick action.

Over 80,000 people are employed in the radio industry at both the local and national levels. This figure includes those who work in commercial radio as well as those employed in broadcasting headquarters (including network offices and public broadcasting stations).

Some people in radio, such as *Disc Jockeys, Radio Announcers,* and *Sports Broadcasters* are prominent around the nation or in their local communities. Their voices and faces (through advertisements, publicity, and public appearances) are familiar to large audiences. Others work behind the scenes, such as *Radio Producers* and *Program Directors. Broadcast Engineers* operate and maintain the equipment. Salespeople selling airtime to advertisers to keep the station profitable are called *Media Buyers.*

In radio, the smallest station may employ only four or five full-time people. Many programs are run by people working on a freelance basis, such as producers, directors, and writers, all employed by a station or a production company from project to project.

Another broadcasting medium is television. In any given hour, a cable or satellite viewer may have a choice of 100 programs; this number is increasing as more specialized cable channels develop original programming, and as digital television makes it possible to receive well over 100 channels in your home. And as we channel-surf with ease, sitting back with our remote controls, thousands of people are hard at work to bring us these programs.

Similar to radio personalities, *Television Announcers* are prominent around the nation or their local community. Their voices and faces become familiar to large audiences. Those working behind the scenes, such as *Television Directors* and *Producers,* are responsible for coordinating all the efforts of production. As with radio, television stations make revenue from advertisements—media buyers sell airtime to keep stations profitable.

Large stations located in metropolitan centers can employ several hundred people, whereas a smaller station in a small city may employ as few as 35 people. Freelance producers, directors, and writers may be used for many projects.

In an attempt to keep the broadcasting world of radio and television in order, the Federal Communications Commission (FCC) was created. The FCC is involved in many aspects of broadcasting, from business matters to the content of programs. The FCC supervises and allocates air space, makes channel assignments, and licenses television stations to applicants who are legally, technically, and financially qualified.

The commission also sets limits on the number of broadcasting stations that a single individual or organization can control. These limits were relaxed, however, in the mid-1980s; broadcasters were no longer required to perform as many public affairs and public service functions, the license renewal process became easier, ownership requirements were loosened, and the process of buying and selling a station became less complicated. The Telecommunications Act of 1996 further lessened ownership restrictions, which resulted in more consolidation among radio and television stations.

Employment in the radio and television broadcasting industry is expected to increase about as fast as the average—about 10 percent—for all other occupations through 2010, according to the U.S. Department of Labor. Competition is the name of the game in the broadcasting industry. The public has many viewing and listening choices, from cable, satellite TV, and conventional television to radio and Web broadcasts. This bodes well for workers as the industry expands and competes for listeners and viewers. However, job seekers should be prepared for stiff competition, since this field is an attractive and popular career choice. Be prepared to work in a smaller market to gain experience; most larger market stations, such as Chicago and New York, prefer to hire experienced workers.

Each article in this book discusses a particular broadcasting occupation in detail. The articles in *Careers in Focus: Broadcasting* appear in Ferguson's *Encyclopedia of Careers and Vocational Guidance*—but have been updated and revised with the latest information from the U.S. Department of Labor and other sources. The **Overview** section is a brief introductory description of the duties and responsibilities of someone in the career. Oftentimes, a career may have a variety of job titles. When this is the case, alternative career titles are presented in this section. The **History** section describes the history of the particular job as it relates to the overall development of its industry or field. **The Job** describes the primary and secondary duties of the job. **Requirements** discusses high school and postsecondary education and training requirements, any certification or licensing necessary, and any other personal requirements for success in the job. **Exploring** offers suggestions on how to gain some experience in or knowledge of the particular job before making a firm educational and financial commitment. The focus is on what can be done while still in high school (or in the early years of college) to gain a better understanding of the job. The **Employers** section gives an overview of typical places of employment for the job. **Starting Out** discusses the best ways to land that first job, be it through the college placement office, newspaper ads, or personal contact. The **Advancement** section describes what kind of career path to expect from the job and how to get there. **Earnings** lists salary ranges and describes the typical fringe benefits. The **Work Environment** section describes the typical surroundings and conditions of employment—whether indoors or outdoors, noisy or quiet, social or independent, and so on. Also discussed are typical hours worked, any seasonal fluctuations, and the stresses and strains of the job. The **Outlook** section summarizes the job in terms of the general economy and industry projections. For the most part, **Outlook** information is obtained from the Bureau of Labor Statistics and is supplemented by information taken from professional associations. Job growth terms follow those used in the *Occupational Outlook Handbook:* Growth described as "much faster than the average" means an increase of 36 percent or more. Growth described as "faster than the average" means an increase of 21 to 35 percent. Growth described as "about as fast as the average" means an increase of 10 to 20 percent. Growth described as "little change or more slowly than the average" means an increase of 0 to 9 percent. "Decline" means a decrease of 1 percent or more. Each article ends with **For More Information,** which lists organizations that can provide career information on training, education, internships, scholarships, and job placement.

Actors

Overview

Actors play parts or roles in dramatic productions on the stage, in motion pictures, or on television or radio. They impersonate, or portray, characters by speech, gesture, song, and dance. There are approximately 105,000 actors employed in the United States.

History

Drama was refined as an art form by the ancient Greeks, who used the stage as a forum for topical themes and stories. The role of actors became more important than in the past, and settings became more realistic with the use of scenery. Playgoing was often a great celebration, a tradition carried on by the Romans. The rise of the Christian church put an end to theater in the sixth century AD, and for several centuries, actors were ostracized from society, surviving as jugglers and jesters.

Drama was reintroduced during the Middle Ages but became more religious in focus. Plays during this period typically centered around biblical themes, and roles were played by priests and other amateurs. This changed with the rediscovery of Greek and Roman plays in the Renaissance. Professional actors and acting troupes toured the countries of Europe, presenting ancient plays or improvising new dramas based on cultural issues and situations of the day. Actors began to take on more prominence in society. In England, actors such as Will Kemp (?-1603?) and Richard Burbage (1567-1619) became known for their roles in the plays of William Shakespeare (1564-1616). In France, Moliere (1622-73) wrote and

Quick Facts

School Subjects
English
Theater/dance
Personal Skills
Artistic
Communication/ideas
Work Environment
Indoors and outdoors
Primarily multiple locations
Minimum Education Level
High school diploma
Salary Range
$7,500 to $25,920 to $1,000,000+
Certification or Licensing
None available
Outlook
Faster than the average

often acted in his own plays. Until the mid-17th century, however, women were banned from the stage, and the roles of women were played by young boys.

By the 18th century, actors could become quite prominent members of society, and plays were often written—or, in the case of Shakespeare's plays, rewritten—to suit a particular actor. Acting styles tended to be highly exaggerated, with elaborate gestures and artificial speech, until David Garrick (1717-79) introduced a more natural style to the stage in the mid-1700s. The first American acting company was established in Williamsburg, Virginia, in 1752, led by Lewis Hallan. In the next century, many actors became stars: famous actors of the time included Edwin Forrest (1806-72), Fanny (1809-93) and Charles Kemble (1775-1854), Edmund Kean (1787-1833), William Charles Macready (1793-1873), and Joseph Jefferson (1829-1905), who was particularly well known for his comedic roles.

Until the late 19th century, stars dominated the stage. But in 1874, George II, Duke of Saxe-Meiningen, formed a theater troupe in which every actor was given equal prominence. This ensemble style influenced others, such as Andre Antoine of France, and gave rise to a new trend in theater called naturalism, which featured far more realistic characters in more realistic settings than before. This style of theater came to dominate the 20th century. It also called for new methods of acting. Konstantin Stanislavsky (1863-1938) of the Moscow Art Theater, who developed an especially influential acting style that was later called method acting, influenced the Group Theater in the United States; one member, Lee Strasberg (1901-82), founded the Actors Studio in New York, which would become an important training ground for many of the great American actors. In the early 20th century, vaudeville and burlesque shows were extremely popular and became the training ground for some of the great comic actors of the century.

By then, developments such as film, radio, and television offered many more acting opportunities than ever before. Many actors honed their skills on the stage and then entered one of these new media, where they could become known throughout the nation and often throughout the world. Both radio and television offered still more acting opportunities in advertisements. The development of sound in film caused many popular actors from the silent era to fade from view, while giving rise to many others. But almost from the beginning, film stars were known for their outrageous salaries and lavish style of living.

In the United States, New York gradually became the center of theater and remains so, although community theater companies abound throughout the country. Hollywood is the recognized center of the motion picture and television industries. Other major production centers are Miami, Chicago, San Francisco, Dallas, and Houston.

The Job

The imitation or basic development of a character for presentation to an audience often seems like a glamorous and fairly easy job. In reality, it is demanding, tiring work requiring a special talent.

The actor must first find a part available in some upcoming production. This may be in a comedy, drama, musical, or opera. Then, having read and studied the part, the actor must audition before the director and other people who have control of the production. This requirement is often waived for established artists. In film and television, actors must also complete screen tests, which are scenes recorded on film, at times performed with other actors, which are later viewed by the director and producer of the film.

If selected for the part, the actor must spend hundreds of hours in rehearsal and must memorize many lines and cues. This is especially true in live theater; in film and television, actors may spend less time in rehearsal and sometimes improvise their lines before the camera, often performing several attempts, or "takes," before the director is satisfied. Television actors often take advantage of teleprompters, which scroll their lines on a screen in front of them while performing. Radio actors generally read from a script, and therefore rehearsal times are usually shorter.

In addition to such mechanical duties, the actor must determine the essence of the character being portrayed and the relation of that character to the overall scheme of the play. Radio actors must be especially skilled in expressing character and emotion through voice alone. In many film and theater roles, actors must also sing and dance and spend additional time rehearsing songs and perfecting the choreography. Some roles require actors to perform various stunts, which can be quite dangerous. Most often, these stunts are performed by specially trained *stunt performers*.

Others work as *stand-ins* or *body doubles*. These actors are chosen for specific features and appear on film in place of the lead actor; this is often the case in films requiring nude or seminude scenes. Many television programs, such as game shows, also feature *models*, who generally assist the host of the program.

Actors in the theater may perform the same part many times a week for weeks, months, and sometimes years. This allows them to develop the role, but it can also become tedious. Actors in films may spend several weeks involved in a production, which often takes place on location, that is, in different parts of the world. Television actors involved in a series, such as a soap opera or a situation comedy, also may play the same role for years, generally in 13-week cycles. For these actors, however, their lines change from week to week and even from day to day, and much time is spent rehearsing their new lines.

While studying and perfecting their craft, many actors work as extras, the nonspeaking characters who appear in the background on screen or stage. Many actors also continue their training. A great deal of an actor's time is spent attending auditions.

Requirements

HIGH SCHOOL
There are no minimum educational requirements to become an actor. However, at least a high school diploma is recommended.

POSTSECONDARY TRAINING
As acting becomes more and more involved with the various facets of our society, a college degree will become more important to those who hope to have an acting career. It is assumed that the actor who has completed a liberal arts program is more capable of understanding the wide variety of roles that are available. Therefore, it is strongly recommended that aspiring actors complete at least a bachelor's degree program in theater or the dramatic arts. In addition, graduate degrees in the fine arts or in drama are nearly always required should the individual decide to teach dramatic arts.

College can also serve to provide acting experience for the hopeful actor. More than 500 colleges and universities throughout the country offer dramatic arts programs and present theatrical performances. Actors and directors recommend that those interested in acting gain as much experience as possible through acting in plays in high school and college or in those offered by community groups. Training beyond college is recommended, especially for actors interested in entering the theater. Joining acting workshops, such as the Actors Studio, can often be highly competitive.

OTHER REQUIREMENTS

Prospective actors will be required not only to have a great talent for acting but also a great determination to succeed in the theater and motion pictures. They must be able to memorize hundreds of lines and should have a good speaking voice. The ability to sing and dance is important for increasing the opportunities for the young actor. Almost all actors, even the biggest stars, are required to audition for a part before they receive the role. In film and television, they will generally complete screen tests to see how they will appear on film. In all fields of acting, a love for acting is a must. It might take many years for an actor to achieve any success, if at all.

Performers on the Broadway stages must be members of the Actors' Equity Association before being cast. While union membership may not always be required, many actors find it advantageous to belong to a union that covers their particular field of performing arts. These organizations include the Actors' Equity Association (stage), Screen Actors Guild or Screen Extras Guild (motion pictures and television films), or American Federation of Television and Radio Artists (TV, recording, and radio). In addition, some actors may benefit from membership in the American Guild of Variety Artists (nightclubs, and so on), American Guild of Musical Artists (opera and ballet), or organizations such as the Hebrew Actors Union or Italian Actors Union for productions in those languages.

Exploring

The best way to explore this career is to participate in school or local theater productions. Even working on the props or lighting crew will provide insight into the field.

Also, attend as many dramatic productions as possible and try to talk with people who either are currently in the theater or have been at one time. They can offer advice to individuals interested in a career in the theater.

Many books, such as *Beginning* (St. Martin's, 1989), by Kenneth Branagh, have been written about acting, not only concerning how to perform but also about the nature of the work, its offerings, advantages, and disadvantages.

Employers

Motion pictures, television, and the stage are the largest fields of employment for actors, with television commercials representing as much as 60 percent of all acting jobs. Most of the opportunities for employment in these fields are either in Los Angeles or in New York. On stage, even the road shows often have their beginning in New York, with the selection of actors conducted there along with rehearsals. However, nearly every city and most communities present local and regional theater productions.

As cable television networks continue to produce more and more of their own programs and films, they will become a major provider of employment for actors. Home video will also continue to create new acting jobs, as will the music video business.

The lowest numbers of actors are employed for stage work. In addition to Broadway shows and regional theater, there are employment opportunities for stage actors in summer stock, at resorts, and on cruise ships.

Starting Out

Probably the best way to enter acting is to start with high school, local, or college productions and to gain as much experience as possible on that level. Very rarely is an inexperienced actor given an opportunity to perform on stage or in film in New York or Hollywood. The field is extremely difficult to enter; the more experience and ability beginners have, however, the greater the possibilities for entrance.

Those venturing to New York or Hollywood are encouraged first to have enough money to support themselves during the long waiting and searching period normally required before a job is found. Most will list themselves with a casting agency that will help them find a part as an extra or a bit player, either in theater or film. These agencies keep names on file along with photographs and a description of the individual's features and experience, and if a part comes along that may be suitable, they contact that person. Very often, however, names are added to their lists only when the number of people in a particular physical category is low. For instance, the agency may not have enough athletic young women on their roster, and if the applicant happens to fit this description, her name is added.

Advancement

New actors will normally start in bit parts and will have only a few lines to speak, if any. The normal procession of advancement would then lead to larger supporting roles and then, in the case of theater, possibly to a role as understudy for one of the main actors. The understudy usually has an opportunity to fill in should the main actor be unable to give a performance. Many film and television actors get their start in commercials or by appearing in government and commercially sponsored public service announcements, films, and programs. Other actors join the afternoon soap operas and continue on to evening programs. Many actors have also gotten their start in on-camera roles such as presenting the weather segment of a local news program. Once an actor has gained experience, he or she may go on to play stronger supporting roles or even leading roles in stage, television, or film productions. From there, an actor may go on to stardom. Only a very small number of actors ever reach that pinnacle, however.

Some actors eventually go into other, related occupations and become drama coaches, drama teachers, producers, stage directors, motion picture directors, television directors, radio directors, stage managers, casting directors, or artist and repertoire managers. Others may combine one or more of these functions while continuing their career as an actor.

Earnings

The wage scale for actors is largely controlled through bargaining agreements reached by various unions in negotiations with producers. These agreements normally control the minimum salaries, hours of work permitted per week, and other conditions of employment. In addition, each artist enters into a separate contract that may provide for higher salaries.

In 2002, the minimum daily salary of any member of the Screen Actors Guild (SAG) in a speaking role was $655, or $2,272 for a five-day workweek. Motion picture actors may also receive additional payments known as residuals as part of their guaranteed salary. Many motion picture actors receive residuals whenever films, TV shows, and TV commercials in which they appear are rerun, sold for TV exhibition, or put on videocassette. Residuals often exceed the actors' original salary and account for about one-third of all actors' income.

A wide range of earnings can be seen when reviewing the Actors' Equity Association's *Annual Report 2000,* which includes a breakdown of average weekly salaries by contract type and location. According to the report, for example, those in "Off Broadway" productions earned an average weekly salary of $642 during the 1999-2000 season. Other average weekly earnings for the same period include: San Francisco Bay area theater, $329; New England area theater, $236; DisneyWorld in Orlando, Florida, $704; and Chicago area theater, $406. The report concludes that the median weekly salary for all contract areas is $457. Most actors do not work 52 weeks per year; in fact, the report notes that of the 38,013 members in good standing, only 16,976 were employed. The majority of those employed, approximately 12,000, had annual earnings ranging from $1 to $15,000.

According to the U.S. Department of Labor, the median yearly earnings of all actors was $25,920 in 2000. The department also reported the lowest paid 10 percent earned less than $12,700 annually, while the highest paid 10 percent made more than $93,620.

The annual earnings of persons in television and movies are affected by frequent periods of unemployment. According to SAG, most of its members earn less than $7,500 a year from acting jobs. Unions offer health, welfare, and pension funds for members working over a set number of weeks a year. Some actors are eligible for paid vacation and sick time, depending on the work contract.

In all fields, well-known actors have salary rates above the minimums, and the salaries of the few top stars are many times higher. Actors in television series may earn tens of thousands of dollars per week, while a few may earn as much as $1 million or more per week. Salaries for these actors vary considerably and are negotiated individually. In film, top stars may earn as much as $20 million per film, and, after receiving a percentage of the gross earned by the film, these stars can earn far, far more.

Until recent years, female film stars tended to earn lower salaries than their male counterparts; the emergence of stars such as Julia Roberts, Jodie Foster, Halle Berry, and others has started to reverse that trend. The average annual earnings for all motion picture actors, however, are usually low for all but the best-known performers because of the periods of unemployment.

Work Environment

Actors work under varying conditions. Those employed in motion pictures may work in air-conditioned studios one week and be on location in a hot desert the next.

Those in stage productions perform under all types of conditions. The number of hours employed per day or week vary, as do the number of weeks employed per year. Stage actors normally perform eight shows per week with any additional performances paid for as overtime. The basic workweek after the show opens is about 36 hours unless major changes in the play are needed. The number of hours worked per week is considerably more before the opening, because of rehearsals. Evening work is a natural part of a stage actor's life. Rehearsals often are held at night and over holidays and weekends. If the play goes on the road, much traveling will be involved.

A number of actors cannot receive unemployment compensation when they are waiting for their next part, primarily because they have not worked enough to meet the minimum eligibility requirements for compensation. Sick leaves and paid vacations are not usually available to the actor. However, union actors who earn the minimum qualifications now receive full medical and health insurance under all the actors' unions. Those who earn health plan benefits for 10 years become eligible for a pension upon retirement. The acting field is very uncertain. Aspirants never know whether they will be able to get into the profession, and, once in, there are uncertainties as to whether the show will be well received and, if not, whether the actors' talent can survive a bad show.

Outlook

Employment in acting is expected to grow faster than the average through 2010, according to the U.S. Department of Labor. There are a number of reasons for this. The growth of satellite and cable television in the past decade has created a demand for more actors, especially as the cable networks produce more and more of their own programs and films. The rise of home video has also created new acting jobs, as more and more films are made strictly for the home video market. Many resorts built in the 1980s and 1990s present their own theatrical productions, providing more job opportunities for actors. Jobs in theater, however, face pressure as the cost of mounting a production rises and as many nonprofit and smaller theaters lose their funding.

Despite the growth in opportunities, there are many more actors than there are roles, and this is likely to remain true for years to come. This is true in all areas of the arts, including radio, television, motion pictures, and theater, and even those who are employed are normally employed during only a small portion of the year. Many actors must supplement their income by working in other areas, such as secretaries, waiters, or taxi drivers, for example. Almost all performers are members of more than one union in order to take advantage of various opportunities as they become available.

It should be recognized that of the 105,000 or so actors in the United States today, an average of only about 16,000 are employed at any one time. Of these, few are able to support themselves on their earnings from acting, and fewer still will ever achieve stardom. Most actors work for many years before becoming known, and most of these do not rise above supporting roles. The vast majority of actors, meanwhile, are still looking for the right break. There are many more applicants in all areas than there are positions. As with most careers in the arts, people enter this career out of a love and desire for acting.

For More Information

The following is a professional union for actors in theater and "live" industrial productions, stage managers, some directors, and choreographers.

ACTORS' EQUITY ASSOCIATION
165 West 46th Street, 15th Floor
New York, NY 10036
Tel: 212-869-8530
Web: http://www.actorsequity.org

This union represents television and radio performers, including actors, announcers, dancers, disc jockeys, newspersons, singers, specialty acts, sportscasters, and stuntpersons.

AMERICAN FEDERATION OF TELEVISION AND RADIO ARTISTS
260 Madison Avenue
New York, NY 10016
Tel: 212-532-0800
Email: aftra@aftra.com
Web: http://www.aftra.com

A directory of theatrical programs may be purchased from NAST. For answers to a number of frequently asked questions concerning education, visit the NAST Web site.

NATIONAL ASSOCIATION OF SCHOOLS OF THEATER (NAST)
11250 Roger Bacon Drive, Suite 21
Reston, VA 20190
Tel: 703-437-0700
Email: info@arts-accredit.org
Web: http://www.arts-accredit.org/nast

The following union provides general information on actors, directors, and producers. Visit the SAG Web site for more information.

SCREEN ACTORS GUILD (SAG)
5757 Wilshire Boulevard
Los Angeles, CA 90036
Tel: 323-954-1600
Web: http://www.sag.com

For information about opportunities in not-for-profit theaters, contact:

THEATRE COMMUNICATIONS GROUP
355 Lexington Avenue
New York, NY 10017
Tel: 212-697-5230
Web: http://www.tcg.org

This site has information for beginners on acting and the acting business.

ACTING WORKSHOP ON-LINE
Web: http://www.redbirdstudio.com/AWOL/acting2.html

Art Directors

Overview

Art directors play a key role in every stage of the creation of an advertisement or ad campaign, from formulating concepts to supervising production. Ultimately, they are responsible for planning and overseeing the presentation of clients' messages in print or on screen-that is, in books, magazines, newspapers, television commercials, posters, and packaging, as well as in film and video and on the World Wide Web.

In publishing, art directors work with artists, photographers, and text editors to develop visual images and generate copy, according to the marketing strategy. They are responsible for evaluating existing illustrations, determining presentation styles and techniques, hiring both staff and freelance talent, working with layouts, and preparing budgets.

In films, videos, and television commercials, art directors set the general look of the visual elements and approve the props, costumes, and models. In addition, they are involved in casting, editing, and selecting the music. In film (motion pictures) and video, the art director is usually an experienced animator or computer/graphic arts designer who supervises the animators.

In sum, art directors are charged with selling to, informing, and educating consumers. They supervise both in-house and off-site staff, handle executive issues, and oversee the entire artistic production process. There are over 147,000 artists and art directors working in the United States.

History

Artists have always been an important part of the creative process. In illustrating the first books, artists painted their subjects by hand using a technique called "illumination," which required putting egg-white tempera on vellum. Each copy of each book had to be printed and illustrated individually, often by the same person.

Printed illustrations first appeared in books in 1461. Through the years, prints were made through lithography, woodblock, and other means of duplicating images. Although making many copies of the same illustration was now possible, publishers still depended on individual artists to create the original works. Text editors usually decided what was to be illustrated and how, while artists commonly supervised the production of the artwork.

The first art directors were probably staff illustrators for book publishers. As the publishing industry grew more complex, with such new technologies as photography and film, art direction evolved into a more supervisory position and became a full-time job. Publishers and advertisers began to need specialists who could acquire and use illustrations. Women's magazines, such as *Vogue* and *Harper's Bazaar,* and photo magazines, such as *National Geographic,* relied so much on illustration that the photo editor and art director began to carry as much power as the text editor.

With the creation of animation, art directors became more indispensable than ever. Animated short films, such as the early Mickey Mouse cartoons, were usually supervised by art directors. Walt Disney, himself, was the art director on many of his early pictures. And as full-length films have moved into animation, the sheer number of illustrations requires more than one art director to oversee the project.

Today's art directors supervise almost every type of visual project produced. Through a variety of methods and media, from television and film to magazines and comic books, art directors communicate ideas by selecting and supervising every element that goes into the finished product.

The Job

Art directors are responsible for all visual aspects of printed or on-screen projects. Overseeing the process of developing visual solutions to a variety of communication problems, the art director helps to establish corporate identities; advertises products and services; enhances books, magazines, newsletters, and other publications; and creates television commercials, film and video productions, and Web sites. Some art directors with experience or knowledge in specific fields specialize in such areas as packaging, exhibitions and displays, or the Internet. But all directors, even those with specialized backgrounds, must be skilled in and knowledgeable about not only design and illustration but also photography, computers, research, and writing, in order to supervise the work of graphic artists, photographers, copywriters, text editors, and other employees.

In print advertising and publishing, art directors may begin with the client's concept or develop one in collaboration with the copywriter and account executive. Once the concept is established, the next step is to decide on the most effective way to communicate it. If there is text, for example, should the art director choose illustrations based on specific text references, or should the illustrations fill in the gaps in the copy? If a piece is being revised, existing illustrations must be reevaluated.

After deciding what needs to be illustrated, art directors must find sources that can create or provide the art. Photo agencies, for example, have photos and illustrations on thousands of different subjects. If, however, the desired illustration does not exist, it may have to be commissioned or designed by one of the staff designers. Commissioning artwork means that the art director contacts a photographer or illustrator and explains what is needed. A price is negotiated, and the artist creates the image specifically for the art director.

Once the illustrations have been secured, they must be presented in an appealing manner. The art director supervises (and may help in the production of) the layout of the piece and presents the final version to the client or creative director. Laying out is the process of figuring out where every image, headline, and block of text will be placed on the page. The size, style, and method of reproduction must all be specifically indicated so that the image is recreated as the director intended it.

In broadcast advertising and film and video, the art director has a wide variety of responsibilities and often interacts with an enormous number of creative professionals. Working with directors and producers, art directors interpret scripts and create or select settings in order to visually convey the story or the message. Ultimately responsible for all visual aspects of the finished product, the art director oversees and channels the talents of set decorators and designers, model makers, location managers, propmasters, construction coordinators, and special effects people. In addition, art directors work with writers, unit production managers, cinematographers, costume designers, and post-production staff, including editors and employees responsible for scoring and titles.

The process of producing a television commercial begins in much the same way that a printed advertising piece is created. The art director may start with the client's concept or create one in-house in collaboration with staff members. Once a concept has been created and the copywriter has generated the corresponding text, the art director sketches a rough storyboard based on the writer's ideas, and the plan is presented for review to the creative director. The next step is to develop a finished storyboard, with larger and more detailed frames (the individual scenes) in color. This storyboard is presented to the client for review and used as a guide for the film director as well.

Technology has been playing an increasingly important role in the art director's job. Most art directors, for example, use a variety of computer software programs, including PageMaker, QuarkXPress, CorelDRAW, FrameMaker, Adobe Illustrator, and Photoshop. Many others create and oversee Web sites for clients and work with other interactive media and materials, including CD-ROM, touch-screens, multidimensional visuals, and new animation programs.

Art directors usually work on more than one project at a time and must be able to keep numerous, unrelated details straight. They often work under pressure of a deadline and yet must remain calm and pleasant when dealing with clients and staff. Because they are supervisors, art directors are often called upon to resolve problems, not only with projects but with employees as well.

Art directors are not entry-level workers. They usually have years of experience working at lower-level jobs in the field before gaining the knowledge needed to supervise projects. Depending on whether they work primarily in publishing or film, art directors have to know how printing presses operate or how film is processed and be familiar with a variety of production techniques in order to understand the wide range of ways that images can be manipulated to meet the needs of a project.

Requirements

HIGH SCHOOL

A college degree is usually a requirement for those interested in becoming art directors; however, in some instances, it is not absolutely necessary. A variety of courses at the high school level will give students interested in pursuing a degree both a taste of college-level offerings and an idea of the skills necessary for art directors on the job. These courses include art, drawing, art history, graphic design, illustration, advertising, and desktop publishing.

In addition, math courses are important. Most of the elements of sizing an image involve calculating percentage reduction or enlargement of the original picture. This must be done with a great degree of accuracy if the overall design is going to work; for example, type size may have to be figured within one thirty-second of an inch for a print project. Errors can be extremely costly and may make the project look sloppy.

Other useful courses that high school students might want to take include business, computing, English, technical drawing, cultural studies, psychology, and social science.

POSTSECONDARY TRAINING

According to the American Institute of Graphic Arts, nine out of 10 artists have a college degree. Among them, six out of 10 have majored in graphic design, and two out of 10 in fine arts. In addition, almost two out of 10 have a master's degree. Along with general two- and four-year colleges and universities, a number of professional art schools offer two-, three-, or four-year programs with such classes as figure drawing, painting, graphic design, and other art courses, as well as classes in art history, writing, business administration, communications, and foreign languages.

Courses in advertising, marketing, photography, filmmaking, set direction, layout, desktop publishing, and fashion are also important for those interested in becoming art directors. Specialized courses, sometimes offered only at professional art schools, that may be particularly helpful for students who want to go into art direction include typography, animation, storyboard, Web site design, and portfolio development.

Because of the rapidly increasing use of computers in design work, it is essential to have a thorough understanding of how computer art and layout programs work. In smaller companies, the art director may be responsible for operating this equipment; in larger companies, a staff person, under the direction of the art director, may use these programs. In either case, the director must know what can be done with the available equipment.

In addition to coursework at the college level, many universities and professional art schools offer graduates or students in their final year a variety of workshop projects, desktop publishing training opportunities, and internships. These programs provide students with opportunities to develop their personal design styles, as well as their portfolios.

OTHER REQUIREMENTS

The work of an art director requires creativity, imagination, curiosity, and a sense of adventure. Art directors must be able to work with specialized materials, such as graphic designs, as well as make presentations on the ideas behind their work.

The ability to work well with various people and organizations is a must for art directors. They must always be up-to-date on new techniques, trends, and attitudes. And because deadlines are a constant part of the work, an ability to handle stress and pressure well is key.

Accuracy and attention to detail are important parts of the job. When the art is done correctly, the public usually pays no notice. But when a project is done badly or sloppily, many people will notice, even if they have no design training. Other requirements for art directors include time management skills and an interest in media and people's motivations and lifestyles.

Exploring

High school students can get an idea of what an art director does by working on the staff of the school newspaper, magazine, or yearbook. It may also be possible to secure a part-time job assisting the advertising director of the local newspaper or to work at an advertising agency.

Developing your own artistic talent is important, and this can be accomplished through self-training (reading books and practicing) or through courses in painting, drawing, or other creative arts. At the very least, you should develop your "creative eye," that is, your ability to develop ideas visually. One way to do this is by familiarizing yourself with great works, such as paintings or highly creative magazine ads, motion pictures, videos, or commercials.

Students can also become members of a variety of advertising clubs around the nation. In addition to keeping members up-to-date on industry trends, such clubs offer job information, resources, and a variety of other benefits.

Employers

A variety of organizations in virtually all industries employ art directors. They might work at advertising agencies, publishing houses, museums, packaging firms, photography studios, marketing and public relations firms, desktop publishing outfits, digital pre-press houses, or printing companies. Art directors who oversee and produce on-screen products often work for film production houses, Web designers, multimedia developers, computer games developers, or television stations.

While companies of all sizes employ art directors, smaller organizations often combine the positions of graphic designer, illustrator, and art director. And although opportunities for art direction can be found all across the nation and abroad, many larger firms in such cities as Chicago, New York, and Los Angeles usually have more openings, as well as higher pay scales, than smaller companies.

Starting Out

Since an art director's job requires a great deal of experience, it is usually not considered an entry-level position. Typically, a person on a career track toward art director is hired as an assistant to an established director. Recent graduates wishing to enter advertising should have a portfolio of their work containing seven to 10 sample ads to demonstrate their understanding of both the business and the media in which they want to work.

Serving as an intern is a good way to get experience and develop skills. Graduates should also consider taking an entry-level job in a publisher's art department to gain initial experience. Either way, aspiring art directors must be willing to acquire their credentials by working on various projects. This may mean working in a variety of areas, such as advertising, marketing, editing, and design.

College publications offer students a chance to gain experience and develop portfolios. In addition, many students are able to do freelance work while still in school, allowing them to make important industry contacts and gain on-the-job experience at the same time.

Advancement

While some may be content upon reaching the position of art director to remain there, many art directors take on even more responsibility within their organizations, such as becoming television directors, starting advertising agencies, creating their own Web sites, developing original multimedia programs, or launching their own magazines.

Many people who get to the position of art director do not advance beyond the title, but move on to work at more prestigious firms. Competition for positions at companies that have national reputations continues to be keen because of the sheer number of talented people interested. At smaller publications or local companies, the competition may be less intense since candidates are competing primarily against others in the local market.

Earnings

The job title of art director can mean many different things, depending on the company at which the director is employed. According to the U.S. Department of Labor, a beginning art director or an art director working at a small firm can expect to make $30,130 or less per year in 2000, with experienced art directors working at larger companies earning more than $109,440. Median annual earnings for art directors employed in the advertising industry (the largest employer of salaried art directors) were $63,510 in

2000. The median annual earnings for art directors working in all industries were $56,880 in 2000. (Again it is important to note that these positions are not entry level; beginning art directors have probably already accumulated several years of experience in the field for which they were paid far less.)

Most companies employing art directors offer insurance benefits, a retirement plan, and other incentives and bonuses.

Work Environment

Art directors usually work in studios or office buildings. While their work areas are ordinarily comfortable, well lit, and ventilated, they often handle glue, paint, ink, and other materials that pose safety hazards and should, therefore, exercise caution.

Art directors at art and design studios and publishing firms usually work a standard 40-hour week. Many, however, work overtime during busy periods in order to meet deadlines. Similarly, directors at film and video operations and at television studios work as many hours as required—usually many more than 40 per week—in order to finish projects according to predetermined schedules.

While art directors work independently, reviewing artwork and reading copy, much time is spent collaborating with and supervising a team of employees, often consisting of copywriters, editors, photographers, graphic artists, and account executives.

Outlook

The extent to which art director positions are in demand, like many other positions, depends on the economy in general; when times are tough, people and businesses spend less, and cutbacks are made. When the economy is healthy, employment prospects for art directors will be favorable. The U.S. Department of Labor predicts about as fast as the average growth in the employment of art directors. One area that shows particularly good promise for growth is the retail industry, since more and more large retail establishments, especially catalog houses, will be employing in-house advertising art directors.

In addition, producers of all kinds of products continually need advertisers to reach their potential customers, and publishers always want some type of illustrations to enhance their books and magazines. Creators of films and videos also need images in order to produce their programs, and people working with new media are increasingly looking for artists and directors to promote new and existing products and services, enhance their Web sites, develop new multimedia programs, and create multidimensional visuals. People who can quickly and creatively generate new concepts and ideas will be in high demand.

On the other side of the coin, the supply of aspiring artists is expected to exceed the number of job openings. As a result, those wishing to enter the field will encounter keen competition, for salaried, staff positions as well as freelance work. And although the Internet is expected to provide many opportunities for artists and art directors, some firms are hiring employees without formal art or design training to operate computer-aided design systems and oversee work.

For More Information

The AAF is the professional advertising association that binds the mutual interests of corporate advertisers, agencies, media companies, suppliers, and academia. For more information, contact:
AMERICAN ADVERTISING FEDERATION (AAF)
1101 Vermont Avenue, NW, Suite 500
Washington, DC 20005-6306
Email: aaf@aaf.org
Web: http://www.aaf.org

This management-oriented national trade organization represents the advertising agency business. For information, contact:
AMERICAN ASSOCIATION OF ADVERTISING AGENCIES
405 Lexington Avenue, 18th Floor
New York, NY 10174-1801
Tel: 212-682-2500
Web: http://www.aaaa.org

The Art Directors Club is an international, nonprofit organization of directors in advertising, graphic design, interactive media, broadcast design, typography, packaging, environmental design, photography, illustration, and related disciplines. For information, contact:

ART DIRECTORS CLUB
106 West 29th Street
New York, NY 10001
Tel: 212-643-1440
Email: messages@adcny.org
Web: http://www.adcny.org

For information on the graphic arts, contact:
GRAPHIC ARTISTS GUILD
90 John Street, Suite 403
New York, NY 10038-3202
Tel: 800-500-2672
Web: http://www.gag.org

Broadcast Engineers

Overview

Broadcast engineers, also referred to as *broadcast technicians,* or broadcast operators, operate and maintain the electronic equipment used to record and transmit the audio for radio signals and the audio and visual images for television signals to the public. They may work in a broadcasting station or assist in broadcasting directly from an outside site as a *field technician.* Approximately 36,000 broadcast engineers work in the United States.

History

At the end of the 19th century, Guglielmo Marconi (1874-1937), an Italian engineer, successfully sent radio waves across a room in his home and helped launch the 20th-century age of mass communication. Marconi quickly realized the potential for his experiments with radio waves. By 1901, he had established the Marconi Wireless Company in England and the United States and soon after successfully transmitted radio signals across the Atlantic Ocean for the first time.

At first, radio signals were used to transmit information and for communication between two points, but eventually the idea was developed that radio could be used for entertainment, and in 1919, the Radio Corporation of America, or RCA, was founded. Families everywhere gathered around their radios to listen to music, drama, comedy, and news programs. Radio became a commercial success, and radio technology advanced, creating the need for skilled engineers to operate the complicated electronic equipment.

In 1933, frequency modulation, or FM, was introduced; originally there had been only amplitude modulation, or AM. This vastly improved the quality of radio broadcasting. At the same time, experimentation was occurring with higher frequency radio waves, and in 1939 at the World's Fair in New York City, RCA demonstrated television.

The effect television had on changing mass communication was as dramatic as the advent of the radio. Technology continued to advance with the introduction of color imaging, which became widely available in 1953. The number of VHF and UHF channels continued to increase; in the 1970s, cable television and subscription television became available, further increasing the amount and variety of programming. Continuing advances in broadcast technology ensure the need for trained engineers who understand and can maintain the highly technical equipment used in television and radio stations.

One of the recent changes in technology that affects broadcast engineers is the switch from analog to digital signals. These changes provide ongoing challenges for television stations.

The Job

Broadcast engineers are responsible for the transmission of radio and television programming, including live and recorded broadcasts. Broadcasts are usually transmitted directly from the station; however, engineers are capable of transmitting signals on location from specially designed, mobile equipment. The specific tasks of the broadcast engineer depend on the size of the television or radio station. In small stations, engineers have a wide variety of responsibilities. Larger stations are able to hire a greater number of engineers and specifically delegate responsibilities to each engineer. In both small and large stations, however, engineers are responsible for the operation, installation, and repair of the equipment.

The *chief engineer* in both radio and television is the head of the entire technical operation and must orchestrate the activities of all the technicians to ensure smooth programming. He or she is also responsible for the budget and must keep abreast of new broadcast communications technology.

Larger stations also have an *assistant chief engineer* who manages the daily activities of the technical crew, controls the maintenance of the electronic equipment, and ensures the performance standards of the station.

Maintenance technicians are directly responsible for the installation, adjustment, and repair of the electronic equipment.

Video technicians usually work in television stations to ensure the quality, brightness, and content of the visual images being recorded and broadcast. They are involved in several different aspects of broadcasting and videotaping television programs. Technicians who are mostly involved with broadcasting programs are often called *video-control technicians.* In live broadcasts using more than one camera, they operate electronic equipment that selects which picture goes to the transmitter for broadcast. They also monitor on-air programs to ensure good picture quality. Technicians mainly involved with taping programs are often called *videotape-recording technicians.* They record performances on videotape using video cameras and tape-recording equipment, then splice together separate scenes into a finished program; they can create special effects by manipulating recording and re-recording equipment. The introduction of robotic cameras, six-foot-tall cameras that stand on two legs, created a need for a new kind of technician called a video-robo technician. *Video-robo technicians* operate the cameras from a control room computer, using joysticks and a video panel to tilt and focus each camera. With the help of new technology, one person can now effectively perform the work of two or three camera operators. Engineers may work with producers, directors, and reporters to put together videotaped material from various sources. These include networks, mobile camera units, and studio productions. Depending on their employer, engineers may be involved in any number of activities related to editing videotapes into a complete program.

Requirements

HIGH SCHOOL

Take as many classes as you can in mathematics, science, computers, and shop, especially electronics. Speech classes will help you hone your abilities to effectively communicate ideas to others.

POSTSECONDARY TRAINING

Positions that are more advanced require a bachelor's degree in broadcast communications or a related field. To become a chief engineer, you should aim for a bachelor's degree in electronics or electrical engineering. Because field technicians also act as announcers on occasion, speech courses and experience as an announcer in a school radio station can be helpful. Seeking education beyond a bachelor's degree will further the possibilities for advancement, although it is not required.

CERTIFICATION OR LICENSING

The Federal Communications Commission licenses and permits are no longer required of broadcast engineers. However, certification from the Society of Broadcast Engineers (SBE) is desirable, and certified engineers consistently earn higher salaries than uncertified engineers. The SBE offers an education scholarship and accepts student members; members receive a newsletter and have access to their job line.

OTHER REQUIREMENTS

Broadcast engineers must have both an aptitude for working with highly technical electronic and computer equipment and minute attention to detail to be successful in the field. Broadcast engineers should enjoy both the technical and artistic aspects of working in the radio or television industry. They should also be able to communicate with a wide range of people with various levels of technical expertise.

Exploring

Experience is necessary to begin a career as a broadcast engineer, and volunteering at a local broadcasting station is an excellent way to gain experience. Many schools have clubs for persons interested in broadcasting. Such clubs sponsor trips to broadcasting facilities, schedule lectures, and provide a place where students can meet others with similar interests. Local television station technicians are usually willing to share their experiences with interested young people. They can be a helpful source of informal career guidance. Visits or tours can be arranged by school officials. Tours allow students to see engineers involved in their work. Most colleges and universities also have radio and television stations where students can gain experience with broadcasting equipment.

Exposure to broadcasting technology also may be obtained through building and operating an amateur, or ham, radio and experimenting with electronic kits. Dexterity and an understanding of home-operated broadcasting equipment will aid in promoting success in education and work experience within the field of broadcasting.

Employers

According to the Federal Communications Commission, there were 13,012 radio stations and 1,686 television stations in the United States in 2001. These stations might be independently operated or owned and operated by a network. Smaller stations in smaller cities are good starting places, but it is in the larger networks and stations in major cities where the higher salaries are found. Some broadcast engineers work outside of the radio and television industries, producing, for example, corporate employee training and sales programs.

Starting Out

In many towns and cities, there are public-access cable television stations and public radio stations where high school students interested in broadcasting and broadcast technology can obtain an internship. An entry-level technician should be flexible about job location; most begin their careers at small stations and with experience may advance to larger-market stations.

Advancement

Entry-level engineers deal exclusively with the operation and maintenance of their assigned equipment; in contrast, a more advanced broadcast engineer directs the activities of entry-level engineers and makes judgments on the quality, strength, and subject of the material being broadcast.

After several years of experience, a broadcast engineer may advance to assistant chief engineer. In this capacity, he or she may direct the daily activities of all of the broadcasting engineers in the station as well as the field engineers broadcasting on location. Advancement to chief engineer usually requires at least a college degree in engineering and many years of experience. A firm grasp of management skills, budget planning, and a thorough knowledge of all aspects of broadcast technology are the requirements to become the chief engineer of a radio or television station.

Earnings

Larger stations usually pay higher wages than smaller stations, and television stations tend to pay more than radio stations. Also, commercial stations generally pay more than public broadcasting stations. The median annual earnings for broadcast technicians were $26,950, according to the U.S. Department of Labor. The department also reported that the lowest paid 10 percent earned less than $13,860 and the highest paid 10 percent earned more than $63,340 during that same period. Experience, job location, and educational background are all factors that influence a person's pay.

Work Environment

Most engineers work in a broadcasting station that is modern and comfortable. The hours can vary; because most broadcasting stations operate 24 hours a day, seven days a week, there are engineers who must work at night, on weekends, and on holidays. Transmitter technicians usually work behind the scenes with little public contact. They work closely with their equipment and as members of a small crew of experts whose closely coordinated efforts produce smooth-running programs. Constant attention to detail and having to make split-second decisions can cause tension. Since broadcasts also occur outside of the broadcasting station on location sites, field technicians may work anywhere and in all kinds of weather.

Outlook

According to the U.S. Department of Labor, the overall employment of broadcast technicians is expected to grow about as fast as the average through 2010. There will be strong competition for jobs in metropolitan areas. In addition, the Department of Labor predicts that a slow growth in the number of new radio and television stations may mean few new job opportunities in the field. Technicians who are able to install transmitters should have better work prospects as television stations switch from their old analog transmitters to digital transmitters. Job openings will also result from the need to replace existing engineers who often leave the industry for other jobs in electronics.

For More Information

For information on its summer internship program, contact:
THE ASSOCIATION OF LOCAL TELEVISION STATIONS
1320 19th Street, NW, Suite 300
Washington, DC 20036
Tel: 202-887-1970
Web: http://www.altv.com

Contact BEA for scholarship information and a list of schools offering degrees in broadcasting. Visit their Web site for useful information about broadcast education and the broadcasting industry.
BROADCAST EDUCATION ASSOCIATION (BEA)
1771 N Street, NW
Washington, DC 20036-2891
Tel: 888-380-7222
Email: beainfo@beaweb.org
Web: http://www.beaweb.org

For broadcast education, support, and scholarship information, contact:
NATIONAL ASSOCIATION OF BROADCASTERS
1771 N Street, NW
Washington, DC 20036-2891
Tel: 202-429-5300
Email: nab@nab.org
Web: http://www.nab.org

For job listings, college programs, and union information, contact:

**NATIONAL ASSOCIATION OF BROADCAST EMPLOYEES
AND TECHNICIANS**

501 3rd Street, NW, 8th Floor
Washington, DC 20001
Tel: 202-434-1254
Email: nabet-cwa@cwa-union.org
Web: http://union.nabetcwa.org

For general information, contact:

NATIONAL ASSOCIATION OF FARM BROADCASTERS

26 Exchange Street, Suite 307
St. Paul, MN 55101
Tel: 651-224-0508
Email: info@nafb.com
Web: http://nafb.com

For a booklet on careers in cable, contact:

NATIONAL CABLE TELEVISION ASSOCIATION

1724 Massachusetts Avenue, NW
Washington, DC 20036-1969
Tel: 202-775-3550
Web: http://www.ncta.com

For scholarship and internship information, contact:

**RADIO-TELEVISION NEWS DIRECTORS ASSOCIATION
AND FOUNDATION**

1600 K Street, NW, Suite 700
Washington, DC 20006-2838
Tel: 202-659-6510
Email: rtnda@rtnda.org
Web: http://www.rtnda.org

*For information on membership, scholarships, and certification,
contact:*

SOCIETY OF BROADCAST ENGINEERS

9247 North Meridian Street, Suite 305
Indianapolis, IN 46260
Tel: 317-846-9000
Web: http://www.sbe.org

Camera Operators

Overview

Camera operators use motion picture cameras and equipment to photograph subjects or material for movies, television programs, or commercials. They usually use 35-millimeter or 16-millimeter cameras or camcorders and a variety of films, lenses, tripods, and filters in their work. Their instructions usually come from cinematographers or directors of photography. Approximately 27,000 camera operators work in the United States.

History

Motion pictures were made as early as 1877, using a series of still photographs to create the illusion of motion. But it was Thomas Edison who, in 1889, produced the first single-unit motion picture camera that set the standard for today.

The motion picture industry blossomed in the United States during the 20th century. With the growth of the television industry and the addition of commercial advertising to television, camera operators became indispensable members of the production crew. Motion picture directors and producers rely on camera operators to create the images on film that the directors and producers envision in their minds. As camera equipment becomes more complex and sophisticated, the camera operator will need to be more proficient at his or her craft.

The Job

Motion picture camera operators may work on feature films in Hollywood or on location elsewhere. Many work on educational films, documentaries, or television programs. The nature of the camera operator's work depends largely on the size of the production crew. If the film is a documentary or short news segment, the camera operator may be responsible for setting up the camera and lighting equipment as well as for supervising the actors during filming. Equipment that camera operators typically use can include cranes, dollies, mounting heads, and different types of lenses and accessories. Often the camera operator is also responsible for maintenance and repair of all of this equipment.

With a larger crew, the camera operator is responsible only for the actual filming. The camera operator may even have a support team of assistants. The *first assistant camera operator* will typically focus on the cameras, make sure cameras are loaded, and confer with lighting specialists. In larger productions, there are also backup cameras and accessories for use if one should malfunction during filming. *Second assistant camera operators* help the first assistant set up scenes to be filmed and assist in the maintenance of the equipment.

Sometimes camera operators must use shoulder-held cameras. This often occurs during the filming of action scenes for television or motion pictures. *Special effects,* or *optical effects, camera operators* photograph the special effects segments for motion pictures and television. They create illusions or effects that can add mood and tone to the motion picture. They usually add fades, dissolves, superimpositions, and other effects to their films at the request of the director of photography, also known as the director of cinematography or the cinematographer.

Brian Fass is a cinematographer/camera assistant in New York City. On a project, he works closely with the other professionals to help establish a visual style for the film. "During the project," he says, "I work on setting up the camera in various positions for coverage of scenes and then lighting each chosen angle." Fass has worked as a camera assistant for the Woody Allen films *Everyone Says I Love You* and *Deconstructing Harry,* as well as the Sidney Lumet film *Gloria.*

Requirements

HIGH SCHOOL

Take classes that will prepare you for the technical aspect of the work—courses in photography, journalism, and media arts should give you some hands-on experience with a camera. Mathematics and science can help you in understanding cameras and filters. You should also take art and art history classes, and other courses that will help you develop appreciation of visual styles.

POSTSECONDARY TRAINING

A college degree is not necessary to get a position as a motion picture camera operator, but a film school can help you expand your network of connections. A bachelor's degree in liberal arts or film studies provides a good background for work in the film industry, but practical experience and industry connections will provide the best opportunities for work. Upon completing an undergraduate program, you may wish to enroll in a master's or master's of fine arts program at a film school. Schools offering well-established programs include the School of Visual Arts in New York, New York University, and the University of Southern California (USC). These schools have film professionals on their faculties and provide a very visible stage for student talent, being located in the two film business hot spots: California and New York. The American Society of Cinematographers (ASC) provides a partial listing of film schools at its Web site. Film school offers overall formal training, providing an education in fundamental skills by working with student productions. Such education is rigorous, but in addition to teaching skills, it provides you with peer groups and a network of contacts with students, faculty, and guest speakers that can be of help after graduation.

OTHER REQUIREMENTS

You must be able to work closely with other members of a film crew and to carefully follow the instructions of the cinematographer and other camera operators. Since lighting is an integral part of filmmaking, you should have a thorough understanding of lighting equipment in order to work quickly and efficiently. In addition to the technical aspects of filmmaking, you should also understand the artistic nature of setting up shots. "I'm dyslexic and have always gravitated toward the visual mediums," Brian Fass says. "I feel that this impairment, along with my love of movies, made me turn toward cinematography."

Exploring

You should join a photography or camera club, or become involved with the media department of your school. You may have the opportunity then to videotape sports events, concerts, and school plays. You can also learn about photography by working in a camera shop. A part-time job in a camera shop will give you a basic understanding of photographic equipment. Some school districts have television stations where students can learn the basics of camera operation. This kind of hands-on experience is invaluable when it comes time to find work in the field. You can also learn about the film industry by reading such publications as *American Cinematographer* and *Cinefex*.

Employers

There are approximately 27,000 television, video, and movie camera operators working in the United States. About 25 percent of these operators are self-employed. The majority of camera operators who are salaried employees work for the film and television industry at TV stations or film studios. Most jobs are found in large, urban areas.

Starting Out

Most entry-level jobs require little formal preparation in photography or camera operation. A college degree is not required by most film or television studios, but you may have to belong to the IATSE Local 600, the union for camera operators. An entry-level job as a camera assistant usually begins with assignments such as setting up or loading film into cameras and adjusting or checking lighting. With experience, the assistant may participate in decisions about what to photograph or how to film a particular scene.

Before you receive any paying jobs, you may have to work for awhile as a volunteer or intern on a film project. You can surf the Internet for postings of openings on film productions, or contact your state's film commission.

Advancement

It usually takes two to four years for a motion picture camera operator to learn the techniques necessary for the job. Those who become proficient in their field, after several years of training, may be able to work on film projects as a *director of photography* (DP). The DP supervises other camera operators, and works more closely with the directors, producers, and actors in the creation of the film. Some camera operators study cinematography part-time while keeping their jobs as camera operators. They may later move to larger studios or command higher salaries.

"I work as an assistant for the money," Brian Fass says, "but hope to jump into work as a DP full time if the jobs come along. I also own my own Aaton XTR camera package which makes me more marketable for DP jobs."

Earnings

Self-employed camera operators typically work on a project-by-project basis and may have periods of unemployment between jobs. Those working on movies may be paid per-day, and their role in the creation of the movie may last anywhere from several weeks to several months. Camera operators who are salaried employees of, for example, a television network, have steady employment. Because of these factors and others, such as area of the country in which the operator works and the size of the employer, salaries vary widely for these professionals. The U.S. Department of Labor reports the median annual earnings of all television, video, and movie camera operators as $27,870 in 2000. The department also reports that the lowest paid 10 percent of operators earned less than $14,130 per year, but at the top end of the pay scale, the highest earning 10 percent made more than $63,690 annually.

Salaried employees usually receive benefits such as health insurance, retirement plans, and vacation days. Those who are self-employed must pay for such extras themselves.

Work Environment

Motion picture camera operators work indoors and outdoors. Most work for motion picture studios or in television broadcasting. During filming, a camera operator may spend several weeks or months on location in another city or country. Most often, the camera operator lives at home and comes to work during regular business hours. Hours can be erratic, however, if the film includes scenes that must be shot at night, or if a deadline must be met by after-hours filming.

Much of the work of a camera operator becomes routine after a few years of experience. Camera operators get used to loading and unloading film, carrying cameras and equipment from trucks or workshops into studios or sets, and filming segments over and over again. The glamour of working on motion pictures or television programs may be diminished by the physically demanding work. Also, the actors, directors, and producers are in the limelight. They often receive credit for the work the camera operators have done.

Many camera operators must be available on short notice. Since motion picture camera operators are generally hired to work on one film at a time, there may be long periods during which a camera operator is not working. Few can make a living as self-employed camera operators.

Motion picture camera operators working on documentary or news productions may work in dangerous places. Sometimes they must work in uncomfortable positions or make adjustments for imperfect lighting conditions. They usually operate their cameras while standing hours at a time. Deadline pressure is also a constant in the camera operator's work. Working for directors or producers who are on tight budgets or strict schedules may be very stressful.

Outlook

Employment for camera operators is expected to increase faster than the average for all occupations through 2010, according to the

U.S. Department of Labor. The use of visual images continues to grow in areas, such as communication, education, entertainment, marketing, and research and development. More businesses will make use of video training films and public relations projects that use film. The entertainment industries are also expanding. However, competition for positions is very fierce. Camera operators work in what is considered a desirable and exciting field, and they must work hard and be aggressive to get good jobs, especially in Los Angeles and New York.

For More Information

To learn more this union (which includes camera operators) and aspects of union membership, visit the IATSE Web site.
INTERNATIONAL ALLIANCE OF THEATRICAL STAGE EMPLOYEES (IATSE)
1430 Broadway, 20th Floor
New York, NY 10018
Tel: 212-730-1770
Web: http://www.iatse.lm.com

To learn about student chapters sponsored by SMPTE, contact:
SOCIETY OF MOTION PICTURE AND TELEVISION ENGINEERS (SMPTE)
595 West Hartsdale Avenue
White Plains, NY 10607
Tel: 914-761-1100
Email: smpte@smpte.org
Web: http://www.smpte.org

Visit this Web site organized by the American Society of Cinematographers for a list of film schools and to learn about the career of cinematographer—the next step on the career ladder for camera operators.
CINEMATOGRAPHER.COM
Web: http://www.cinematographer.com

Cartoonists and Animators

Overview

Cartoonists and animators are illustrators who draw pictures and cartoons to amuse, educate, and persuade people.

The Job

Cartoonists draw illustrations for newspapers, books, magazines, greeting cards, movies, television shows, civic organizations, and private businesses. Cartoons most often are associated with newspaper comics or with children's television, but they are also used to highlight and interpret information in publications as well as in advertising.

Quick Facts

School Subjects
Art
Computer science
Personal Skills
Artistic
Communication/ideas
Work Environment
Primarily indoors
Primarily one location
Minimum Education Level
High school diploma
Salary Range
$10,400 to $31,190 to $70,560+
Certification or Licensing
None available
Outlook
About as fast as the average

Whatever their individual specialty, cartoonists and animators translate ideas onto paper or film in order to communicate these ideas to an audience. Sometimes the ideas are original; at other times, they are directly related to the news of the day, to the content of a magazine article, or to a new product. After cartoonists come up with ideas, they discuss them with their employers, who include editors, producers, and creative directors at advertising agencies. Next, cartoonists sketch drawings and submit these for approval. Employers may suggest changes, which the cartoonists then make. Cartoonists use a variety of art materials, including pens, pencils, markers, crayons, paints, transparent washes, and shading sheets. They may draw on paper, acetate, or bristol board.

Animators are relying increasingly on computers in various areas of production. Computers are used to color animation art, whereas formerly, every frame was painted by hand. Computers also help animators create special effects or even entire films.

Comic strip artists tell jokes or short stories with a series of pictures. Each picture is called a frame or a panel, and each frame usually includes words as well as drawings. *Comic book artists* also tell stories with their drawings, but their stories are longer, and they are not necessarily meant to be funny.

Animators, or *motion cartoonists,* also draw individual pictures, but they must draw many more for a moving cartoon. Each picture varies only slightly from the ones before and after it in a series. When these drawings are photographed in sequence to make a film and then the film is projected at high speed, the cartoon images appear to be moving. Animators today also work a great deal with computers.

Other people who work in animation are *prop designers,* who create objects used in animated films, and *layout artists,* who visualize and create the world that cartoon characters inhabit.

Editorial cartoonists comment on society by drawing pictures with messages that are usually funny, but which often have a satirical edge. Their drawings often depict famous politicians. *Portraitists* are cartoonists who specialize in drawing caricatures. Caricatures are pictures that exaggerate someone's prominent features, such as a large nose, to make them recognizable to the public. Most editorial cartoonists are also talented portraitists.

Storyboard artists work in film and television production as well as at advertising agencies. They draw cartoons or sketches that give a client an idea of what a scene or television commercial will look like before it is produced. If the director or advertising client likes the idea, the actions represented by cartoons in the storyboard will be reproduced by actors on film.

Requirements

HIGH SCHOOL

If you are interested in becoming a cartoonist or animator, you should, of course, study art in high school in addition to following a well-rounded course of study. To comment insightfully on contemporary life, it is useful to study political science, history, and social studies. English and communications classes will also help you to become a better communicator.

POSTSECONDARY TRAINING

Cartoonists and animators need not have a college degree, but some art training is usually expected by employers. Animators must attend art school to learn specific technical skills. Training in computers in addition to art can be especially valuable.

OTHER REQUIREMENTS

Cartoonists and animators must be creative. In addition to having artistic talent, they must generate ideas, although it is not unusual for cartoonists to collaborate with writers for ideas. Whether they create cartoon strips or advertising campaigns, they must be able to come up with concepts and images to which the public will respond. They must have a good sense of humor and an observant eye to detect people's distinguishing characteristics and society's interesting attributes or incongruities.

Cartoonists and animators need to be flexible. Because their art is commercial, they must be willing to accommodate their employers' desires if they are to build a broad clientele and earn a decent living. They must be able to take suggestions and rejections gracefully.

Exploring

If you are interested in becoming a cartoonist or an animator, you should submit your drawings to your school paper. You also might want to draw posters to publicize activities, such as sporting events, dances, and meetings.

Scholarship assistance for art students is available from some sources. For example, the Society of Illustrators awards some 125 scholarships annually to student artists from any field. Students do not apply directly; rather, they are selected and given application materials by their instructors. The International Animated Film Society offers scholarships to high school seniors.

Employers

Employers of cartoonists and animators include editors, producers, creative directors at advertising agencies, comics syndicates, newspapers, movie studios, and television networks. In addition, a number of these artists are self-employed, working on a freelance basis.

Starting Out

A few places, such as the Walt Disney studios, offer apprenticeships. To enter these programs, applicants must have attended an accredited art school for two or three years.

Formal entry-level positions for cartoonists and animators are rare, but there are several ways for artists to enter the cartooning field. Most cartoonists and animators begin by working piecemeal, selling cartoons to small publications, such as community newspapers, that buy freelance cartoons. Others assemble a portfolio of their best work and apply to publishers or the art departments of advertising agencies. In order to become established, cartoonists and animators should be willing to work for what equals less than minimum wage.

Advancement

Cartoonists' success, like that of other artists, depends upon how much the public likes their work. Very successful cartoonists and animators work for prestigious clients at the best wages; some become well known to the public.

Earnings

Freelance cartoonists may earn anywhere from $100 to $1,200 or more per drawing, but top dollar generally goes only for big, full-color projects such as magazine cover illustrations. Most cartoonists and animators average from $200 to $1,500 a week ($10,400 to $78,000 per year), although syndicated cartoonists on commission can earn much more. Salaries depend upon the work performed. Cel painters, as listed in a salary survey conducted by *Animation World*, start at about $750 a week; animation checkers,

$930 a week; story sketchers, $1,500 weekly. According to *U.S. News & World Report*, animators, depending on their experience, can earn from $800 to $1,800 a week. Top animators can command weekly fees of about $6,500 or more. The U.S. Department of Labor reports a median annual income of $41,130 for salaried multimedia artists and animators in 2000. Incomes ranged from the lowest paid 10 percent earning less than $23,740 to the highest paid 10 percent making more than $70,560 that same year. Comic strip artists are usually paid according to the number of publications that carry their strip. Although the Department of Labor does not give specific information regarding cartoonists' earnings, it does note that the median earnings for salaried fine artists were $31,190 in 2000. Salaried cartoonists, who are related workers, may have earnings similar to this figure.

Self-employed artists do not receive fringe benefits such as paid vacations, sick leave, health insurance, or pension benefits. Those who are salaried employees of companies, agencies, newspapers and the like do typically receive these fringe benefits.

Work Environment

Most cartoonists and animators work in big cities where employers such as television studios, magazine publishers, and advertising agencies are located. They generally work in comfortable environments, at drafting tables or drawing boards with good light. Staff cartoonists work a regular 40-hour workweek but may occasionally be expected to work evenings and weekends to meet deadlines. Freelance cartoonists have erratic schedules, and the number of hours they work may depend on how much money they want to earn or how much work they can find. They often work evenings and weekends, but are not required to be at work during regular office hours.

Cartoonists and animators can be frustrated by employers who curtail their creativity, asking them to follow instructions that are contrary to what they would most like to do. Many freelance cartoonists spend a lot of time working alone at home, but cartoonists have more opportunities to interact with other people than do most working artists.

Outlook

Employment for artists and related workers is expected to grow at a rate about as fast as the average through 2010, according to the U.S. Department of Labor. Because so many creative and talented people are drawn to this field, however, competition for jobs will be strong.

Cartoons are not just for children anymore. Much of the animation today is geared for an adult audience. Interactive games, animated films, network and cable television, and the Internet are among the many employment sources for talented cartoonists and animators. More than half of all visual artists are self-employed, but freelance work can be hard to come by and many freelancers earn little until they acquire experience and establish a good reputation. Competition for work will be keen; those with an undergraduate or advanced degree in art or film will be in demand. Experience in action drawing and computers is a must.

The growing trend of sophisticated special effects in motion pictures should create opportunities at industry effects houses such as Sony Pictures Image Works, DreamQuest, Industrial Light & Magic, and DreamWorks.

For More Information

For membership and scholarship information, contact:
INTERNATIONAL ANIMATED FILM SOCIETY
721 South Victory Boulevard
Burbank, CA 91502
Email: info@asifa-hollywood.org
Web: http://www.asifa-hollywood.org

For an art school directory, a scholarship guide, or general information, contact:
NATIONAL ART EDUCATION ASSOCIATION
1916 Association Drive
Reston, VA 20191-1590
Tel: 703-860-8000
Email: naea@dgs.dgsys.com
Web: http://www.naea-reston.org

For education and career information, contact:
NATIONAL CARTOONISTS SOCIETY
PO Box 713
Suffield, CT 06078
Web: http://www.reuben.org

For scholarship information for qualified students in art school, have your instructor contact:
SOCIETY OF ILLUSTRATORS
128 East 63rd Street
New York, NY 10021-7303
Web: http://www.societyillustrators.org

Disc Jockeys

Quick Facts

School Subjects
English
Speech

Personal Skills
Communication/ideas

Work Environment
Primarily indoors
Primarily one location

Minimum Education Level
Some postsecondary training

Salary Range
$7,000 to $31,251 to $160,000+

Certification or Licensing
None available

Outlook
Decline

Overview

Disc jockeys play recorded music on radio or during parties, dances, and special events. On the radio, they intersperse the music with a variety of advertising material and informal commentary. They may also perform such public services as announcing the time, the weather forecast, or important news. Interviewing guests and making public service announcements may also be part of the disc jockey's work. There are about 50,000 disc jockeys in the United States.

History

Guglielmo Marconi (1874-1937), a young Italian engineer, first transmitted a radio signal in his home in 1895. Radio developed rapidly as people began to comprehend its tremendous possibilities. The stations KDKA in Pittsburgh and WWJ in Detroit began broadcasting in 1920. Within 10 years, there were radio stations in all the major cities in the United States and broadcasting had become big business. The National Broadcasting Company became the first network in 1926 when it linked together 25 stations across the country. The Columbia Broadcasting System was organized the following year. In 1934, the Mutual Broadcasting Company was founded. The years between 1930 and 1950 may be considered the zenith years for the radio industry. With the coming of television, radio broadcasting took second place in importance as entertainment for the home, but radio's commercial and communications value should not be underestimated.

The first major contemporary disc jockey in the United States was Alan Freed (1921-65), who worked in the 1950s on WINS radio in New York. In 1957, his rock-and-roll stage shows at the Paramount Theater made front-page news in *The New York Times* because of the huge crowds they attracted. The title "disc jockey" arose when most music was recorded on conventional flat records, or discs.

Today, much of the recorded music used in commercial radio stations is on magnetic tape or compact disc. The disc jockey personalities are still very much a part of the radio station's image, with major players commanding salaries at the top of the range.

The Job

Disc jockeys serve as a bridge between the music itself and the listener. They also perform such public services as announcing the time, the weather forecast, or important news. Working at a radio station can be a lonely job, since often the disc jockey is the only person in the studio. But because their job is to maintain the good spirits of their audience and attract new listeners, disc jockeys must possess the ability to sound relaxed and cheerful.

Dave Wineland is a disc jockey at WRZQ 107.3 in Columbus, Indiana. He covers the popular 5:30 to 10 AM morning shift that many commuters listen to on their way to work. Like many disc jockeys, his duties extend beyond on-the-air announcements. He works as production director at the station and writes and produces many of the commercials and promotion announcements. "I spend a lot of time in the production room," says Wineland, who also delegates some of the production duties to other disc jockeys on the staff.

Unlike the more conventional radio or television announcer, the disc jockey is not bound by a written script. Except for the commercial announcements, which must be read as written, the disc jockey's statements are usually spontaneous. Disc jockeys are not usually required to play a musical selection to the end; they may fade out a record when it interferes with a predetermined schedule for commercials, news, time checks, or weather reports. Disc jockeys are not always free to play what they want; at some radio stations, especially the larger ones, the program director or the music director makes the decisions about the music that will be played.

And while some stations may encourage their disc jockeys to talk, others emphasize music over commentary and restrict the amount of a DJ's ad-libbing.

Disc jockeys should be levelheaded and able to react calmly even in the face of a crisis. Many unexpected circumstances can arise that demand the skill of quick thinking. For example, if guests who are to appear on a program either do not arrive or become too nervous to go on the air, the disc jockey must fill the airtime. He or she must also smooth over a breakdown in equipment or some other technical difficulty.

Many disc jockeys have become well-known public personalities in broadcasting; they may participate in community activities and public events.

Disc jockeys who work at parties and other special events usually work on a part-time basis. They are often called *party DJs.* A DJ who works for a supplying company receives training, equipment, music, and job assignments from the company. Self-employed DJs must provide everything they need themselves. Party DJs have more contact with people than radio DJs, so they must be personable and patient with clients.

Requirements

HIGH SCHOOL
In high school, you can start to prepare for a career as a disc jockey. A good knowledge of the English language, correct pronunciation, and diction are important. High school English classes as well as speech classes are helpful in getting a good familiarity with the language. Extracurricular activities such as debating and theater will also help with learning good pronunciation and projection.

Many high schools have radio stations on site where students can work as disc jockeys, production managers, or technicians. This experience can be a good starting point to learn more about the field. Dave Wineland's first radio job was at the radio station at Carmel High School in Indianapolis.

POSTSECONDARY TRAINING
Although there are no formal educational requirements for becoming a disc jockey, many large stations prefer applicants with some

college education. Some schools train students for broadcasting, but such training will not necessarily improve the chances of an applicants' getting a job at a radio station.

Students interested in becoming a disc jockey and advancing to other broadcasting positions should attend a school that will train them to become an announcer. There are some private broadcasting schools that offer good courses, but others are poor; students should get references from the school or the local Better Business Bureau before taking classes.

Like many disc jockeys today, Wineland has a college degree. He earned a degree in telecommunications from Ball State University.

Candidates may also apply for any job at a radio station and work their way up. Competition for disc jockey positions is strong. Although there may not be any specific training program required by prospective employers, station officials pay particular attention to taped auditions of the applicant. Companies that hire DJs for parties will often train them; experience is not always necessary if the applicant has a suitable personality.

OTHER REQUIREMENTS

Union membership may be required for employment with large stations in major cities and is a necessity with the networks. The largest talent union is the American Federation of Television and Radio Artists. Most small stations, however, are nonunion.

Exploring

If becoming a disc jockey sounds interesting, you might try to get a summer job at a radio station. Although you will probably not have any opportunity to broadcast, you may be able to judge whether or not that kind of work appeals to you as a career.

Take advantage of any opportunity you get to speak or perform before an audience. Appearing as a speaker or a performer can help you decide whether or not you have the necessary stage presence for a career on the air.

Many colleges and universities have their own radio stations and offer courses in radio. Students gain valuable experience working at college-owned stations. Some radio stations offer students financial assistance and on-the-job training in the form of internships and co-op work programs, as well as scholarships and fellowships.

Employers

There has been a steady growth in the number of radio stations in the United States. According to 2001 statistics from the National Association of Broadcasters, the United States alone has 13,012 radio stations.

Radio is a 24-hour-a-day, seven-day-a-week medium, so there are many slots to fill. Most of these stations are small stations where disc jockeys are required to perform many other duties for a lower salary than at larger stations in bigger metropolitan areas.

Due to the Telecommunications Act of 1996, companies can own an unlimited number of radio stations nationwide with an eight-station limit within one market area, depending on the size of the market. When this legislation took effect, mergers and acquisitions changed the face of the radio industry. So, while the pool of employers is smaller, the number of stations continues to rise.

Starting Out

One way to enter this field is to apply for an entry-level job rather than a job as a disc jockey. It is also advisable to start at a small local station. As opportunities arise, DJs commonly move from one station to another.

While still a high school student, Dave Wineland applied for a position at his local radio station in Monticello, Indiana. "I was willing to work long hours for low pay," he says, acknowledging that starting out in radio can require some sacrifices. However, on-air experience is a must.

An announcer is employed only after an audition. Audition material should be selected carefully to show the prospective employer the range of the applicant's abilities. A prospective DJ should practice talking aloud, alone, then make a tape of him- or herself with five to seven minutes of material to send to radio stations. The tape should include a news story, two 60-second commercials, and a sample of the applicant introducing and coming out of a record. (Tapes should not include the whole song, just the first and final few seconds, with the aspiring DJ introducing and finishing the music; this is called "telescoping.") In addition to presenting prepared materials, applicants may also be asked to read material that they have not seen previously. This may be a commercial, news release, dramatic selection, or poem.

Advancement

Most successful disc jockeys advance from small stations to large ones. The typical experienced disc jockey will have held several jobs at different types of stations.

Some careers lead from being a disc jockey to other types of radio or television work. More people are employed in sales, promotion, and planning than in performing, and they are often paid more than disc jockeys.

Earnings

The salary range for disc jockeys is extremely broad with a low of $7,000 and a high of $100,000. The average salary in the late 1990s was $31,251, according to a survey conducted by the National Association of Broadcasters and the Broadcast Cable Financial Management Association.

Smaller market areas and smaller stations fall closer to the bottom of the range, while the top markets and top-rated stations offer disc jockeys higher salaries.

In large markets such as Chicago, earnings can range depending on broadcast time. According to a survey by the accounting and consulting firm of Hungerford, Aldrin, Nichols and Carter, morning radio announcers made as much as $160,000 in 1999. Those that work later in the day earn considerably less. Afternoon announcers made an average of $80,000 a year, and evening announcers earned closer to $60,000. In the same report, overnight and weekend disc jockeys earned $40,000 and $27,000, respectively.

Benefits for disc jockeys vary according to the size of the market and station. However, vacation and sick time is somewhat limited because the medium requires that radio personalities be on the air nearly every day.

Work Environment

Work in radio stations is usually very pleasant. Almost all stations are housed in modern facilities. Temperature and dust control are important factors in the proper maintenance of technical electronic equipment, and people who work around such machinery benefit from the precautions taken to preserve it.

The work can be demanding. It requires that every activity or comment on the air begin and end exactly on time. This can be difficult, especially when the disc jockey has to handle news, commercials, music, weather, and guests within a certain time frame. It takes a lot of skill to work the controls, watch the clock, select music, talk with a caller or guest, read reports, and entertain the audience; often several of these tasks must be performed simultaneously. A disc jockey must be able to plan ahead and stay alert so that when one song ends he or she is ready with the next song or with a scheduled commercial.

Because radio audiences listen to disc jockeys who play the music they like and talk about the things that interest them, disc jockeys must always be aware of pleasing their audience. If listeners begin switching stations, ratings go down and disc jockeys can lose their jobs.

Disc jockeys do not always have job security; if the owner or manager of a radio station changes, the disc jockey may lose his or her job. The consolidation of radio stations to form larger, cost-efficient stations has caused some employees to lose their jobs.

Disc jockeys usually work a 40-hour week, but they may work irregular hours. They may have to report for work at a very early hour in the morning. Sometimes they will be free during the daytime hours, but will have to work late into the night. Some radio stations operate on a 24-hour basis. All-night announcers may be alone in the station during their working hours.

The disc jockey who stays with a station for a period of time becomes a well-known personality in the community. Such celebrities are sought after as participants in community activities and may be recognized on the street.

Disc jockeys who work at parties and other events work in a variety of settings. They generally have more freedom to choose music selections but little opportunity to ad-lib. Their work is primarily on evenings and weekends.

Outlook

According to the National Association of Broadcasters, radio reaches 77 percent of people over the age of 12 everyday. Despite radio's popularity, the *Occupational Outlook Handbook* projects that employment of announcers will decline slightly through 2010. Due to this decline, competition for jobs will be great in an already competitive field.

While small stations will still hire beginners, on-air experience will be increasingly important. Another area where job seekers can push ahead of the competition is in specialization. Knowledge of specific areas such as business, consumer, and health news may be advantageous.

While on-air radio personalities are not necessarily affected by economic downturns, mergers and changes in the industry can affect employment. If a radio station has to make cuts due to the economy, it is most likely to do so in the behind-the-scenes area, which means that the disc jockeys who remain may face a further diversity in their duties.

For More Information

For a list of schools offering degrees in broadcasting as well as scholarship information, contact:
BROADCAST EDUCATION ASSOCIATION
1771 N Street, NW
Washington, DC 20036-2891
Tel: 202-429-5354
Web: http://www.beaweb.org

For broadcast education, support, and scholarship information, contact:

NATIONAL ASSOCIATION OF BROADCASTERS
1771 N Street, NW
Washington, DC 20036
Tel: 202-429-5300
Email: nab@nab.org
Web: http://www.nab.org

For college programs and union information, contact:

NATIONAL ASSOCIATION OF BROADCAST EMPLOYEES AND TECHNICIANS
501 Third Street, NW, 8th Floor
Washington, DC 20001
Tel: 202-434-1254
Email: nabet-cwa@cwa-union.org
Web: http://nabetcwa.org

For information on student membership, contact:

NATIONAL ASSOCIATION OF FARM BROADCASTERS
26 Exchange Street, Suite 307
St. Paul, MN 55101
Tel: 651-224-0508
Email: info@nafb.com
Web: http://nafb.com

For scholarship and internship information, contact:

RADIO-TELEVISION NEWS DIRECTORS ASSOCIATION AND FOUNDATION
1600 K Street, NW, Suite 700
Washington, DC 20006-2838
Tel: 202-659-6510
Email: rtnda@rtnda.org
Web: http://www.rtnda.org

Lighting Technicians

Overview

Lighting technicians set up and control lighting equipment for television broadcasts, taped television shows, motion pictures, and video productions. They begin by consulting with the production director and technical director to determine the types of lighting and special effects that are needed. Working with spot and flood lights, mercury-vapor lamps, white and colored lights, reflectors (mainly employed out-of-doors), and a large array of dimming, masking, and switching controls, they light scenes to be broadcast or recorded.

Quick Facts

School Subjects
Mathematics
Technical/shop
Personal Skills
Following instructions
Mechanical/manipulative
Work Environment
Indoors and outdoors
Primarily multiple locations
Minimum Education Level
Some postsecondary training
Salary Range
$16.50 to $20.50 to $22.50 an hour
Certification or Licensing
None available
Outlook
About as fast as the average

History

For centuries before the arrival of electric lights, theaters used candles and oil lamps to make the action on an indoor stage visible. The effects produced were necessarily limited by the lack of technology. In 1879, Thomas A. Edison developed a practical electric light bulb by removing most of the oxygen from a glass bulb and then sending current through a carbon filament inside—producing a light that would not burn out. With the arrival of electric lights, it was only a short time before theater lighting became more sophisticated; spotlights and various lighting filters were put to use, and specialists in lighting emerged.

The manipulation of light and shadow is one of the basic principles of filmmaking. This was particularly the case during the era of the silent film; without sound, filmmakers relied upon image to tell their stories. Lighting professionals learned how to make the illusion complete; through expert lighting, cardboard backdrops could substitute for the outdoors, actors could change appearance, and cheaply constructed costumes could look extravagant. Lighting technicians were the first visual effects masters, using tricks with light to achieve realism. As film techniques and equipment evolved, lighting technicians worked with cinematographers and directors to create the dark recesses and gritty streets of the film noir, the lavish spectacle of the movie musical, and the sweeping plains of the American western, often within the confines of a studio. In the 1960s and 1970s, however, a new movie realism called for lighting technicians to expose, with uneven lighting and weak light sources, all the imperfections they'd been covering up before. Today, in the era of the special effects blockbuster, lighting technicians have gone back to their roots, using light and advanced equipment to make model cities, planets, and monsters look real.

The Job

Whenever a movie or television show is filmed, the location must be well lit, whether indoors in a studio or outdoors on location. Without proper lighting, the cameras would not be able to film properly, and the show would be difficult to watch. Lighting technicians set up and control the lighting equipment for movie and television productions. These technicians are sometimes known as *assistant chief set electricians* or *lights operators.*

When beginning a project, lighting technicians consult with the director to determine the lighting effects needed; then they arrange the lighting equipment and plan the light-switching sequence that will achieve the desired effects. For example, if the script calls for sunshine to be streaming in through a window, they will set up lights to produce this effect. Other effects they may be asked to produce include lightning, the flash from an explosion, or the soft glow of a candle-lit room.

For a television series, which uses a similar format for each broadcast, the lights often remain in one fixed position for every show. For a one-time production, such as a scene in a movie, the lights have to be physically set up according to the particular scene.

During filming, lighting technicians follow a script that they have marked or follow instructions from the technical director. The script tells them which lighting effects are needed at every point in the filming. In a television studio, lighting technicians watch a monitor screen to check the lighting effects. If necessary, they may alter the lighting as the scene progresses by adjusting controls in the control room.

Broadcasts from indoor settings require carrying and setting up portable lights. In small television stations, this work may be done by the camera operator or an assistant. In a large station, or in any big movie or television production, a lighting technician may supervise several assistants as they set up the lights.

Even outdoor scenes require lighting, especially to remove shadows from people's faces. For outdoor scenes in bad weather or on rough terrain, it may be a difficult task to secure the lighting apparatus firmly so that it is out of the way, stable, and protected. During a scene, whether broadcast live or recorded on film, lighting technicians must be able to concentrate on the lighting of the scene and must be able to make quick, sure decisions about lighting changes.

There are different positions within this field, depending on experience. A lighting technician can move up into the position of *best boy* (the term applies to both genders). This person assists the *chief lighting technician,* or *gaffer.* The gaffer is the head of the lighting department and hires the lighting crew. Gaffers must be sure the filmed scene looks the way the director and the director of photography want it to look. They must diagram each scene to be filmed and determine where to position each light and decide what kinds of lights will work best for each particular scene. Gaffers must be observant, noticing dark and bright spots and correcting their light levels before filming takes place.

Requirements

HIGH SCHOOL

You should learn as much as possible about electronics. Physics, mathematics, and any shop courses that introduce electronics equipment provide a good background. You should also take courses that will help you develop computer skills needed for operating lighting and sound equipment. Composition or technical writing courses can give you the writing skills you'll need to communicate ideas to other technicians.

POSTSECONDARY TRAINING

There is strong competition for broadcast and motion picture technician positions, and, in general, only well-prepared technicians get good jobs. You should attend a two-year postsecondary training program in electronics and broadcast technology, especially if you hope to advance to a supervisory position. Film schools also offer useful degrees in production, as do theater degrees. For a position as a chief engineer, a bachelor's degree is usually required.

OTHER REQUIREMENTS

Setting up lights can be heavy work, especially when lighting a large movie set. You should be able to handle heavy lights on stands and work with suspended lights while on a ladder. Repairs such as changing light bulbs or replacing worn wires are sometimes necessary. You should be able to work with electricians' hand tools (screwdrivers, pliers, and so forth) and be comfortable working with electricity. You should also be dependable and capable of working as part of a team. Communications skills, both listening and speaking, are necessary when working with a director and with assistant light technicians.

Exploring

Valuable learning experiences for prospective lighting technicians include working on the lighting for a school stage production, building a radio from a kit, or a summer job in an appliance or TV repair shop. High school shop or vocational teachers may be able to arrange a presentation by a qualified lighting technician.

You can also learn a lot about the technical side of production by operating a camera for your school's journalism or media department. Videotaping a play, concert, or sporting event will give you additional insight into production work. You may also have the opportunity to intern or volunteer with a local technical crew for a film or TV production. Check the Internet for production schedules, or volunteer to work for your state's film commission where you'll hear about area projects.

Employers

Many lighting technicians work on a freelance basis, taking on film, TV, and commercial projects as they come along. Technicians can find full-time work with large theater companies and television broadcast stations, or any organization, such as a museum or sports arena, that requires special lighting. Lighting technicians also work for video production companies.

Starting Out

The best way to get experience is to find a position as an intern. Offering to work for a production for course credit or experience instead of pay will enable you to learn about the job and to establish valuable connections. Most people interested in film and television enter the industry as production assistants. These positions are often unpaid and require a great deal of time and work with little reward. However, production assistants have the opportunity to network with people in the industry. They get to speak to lighting technicians and to see them at work. Once they've worked on a few productions, and have learned many of the basics of lighting, they can negotiate for paid positions on future projects.

Advancement

An experienced lighting technician will be able to move up into the position of best boy. With a few more years experience working under different gaffers on diverse projects, the technician may move into the position of gaffer or chief lighting technician. Gaffers command greater salaries as they gain experience working with as many cinematographers as possible. Many experienced technicians join the International Alliance of Theater Stage Engineers (IATSE) as lighting technicians or studio mechanics; union membership is required for work on major productions.

Some lighting technicians go on to work as cinematographers, or to make their own films or television movies.

Earnings

Salaries for lighting technicians vary according to the technician's experience. Annual income is also determined by the number of projects a technician is hired for; the most experienced technicians can work year-round on a variety of projects, while those starting out may go weeks without work. According to IATSE, the minimum hourly pay for unionized gaffers was $22.50 in 2000. Unionized best boys and other lighting technicians earned at least $16.50 to $20.50 an hour. Experienced technicians can negotiate for much higher wages. Union members are also entitled to certain health and retirement benefits.

Work Environment

Lighting technicians employed in television normally work a 40-hour week and change jobs only as their experience makes advancement possible. Technicians employed in the motion picture industry are often employed for one production at a time and thus may work less regularly and under a more challenging variety of conditions than light technicians in television studio work.

Technicians often work long days, especially when a film is on a tight schedule or when news teams are covering a late-breaking story. Technicians may travel a good deal to be on location for filming. They work both indoors in studios and outdoors on location, under a variety of weather conditions.

Outlook

As long as the movie and television industries continue to grow, opportunities will remain available for people who wish to become lighting technicians. With the expansion of the cable market, lighting technicians may find work in more than one area. However, persistence and hard work are required in order to secure a good job in film or television.

The increasing use of visual effects and computer generated imagery (CGI) will likely have an impact on the work of lighting technicians. Through computer programs, filmmakers and editors can adjust lighting themselves; however, live-action shots are still integral to the filmmaking process, and will remain so for some time. To get the initial shots of a film will require sophisticated lighting equipment and trained technicians. Lighting technicians often have to know about the assembly and operation of more pieces of equipment than anyone else working on a production. Equipment will become more compact and mobile, making the technician's job easier.

For More Information

For information about colleges with film and television programs of study, and to read interviews with filmmakers, visit the AFI Web site:
AMERICAN FILM INSTITUTE (AFI)
2021 North Western Avenue
Los Angeles, CA 90027
Tel: 323-856-7600
Web: http://www.afi.com

Visit the ASC Web site for a great deal of valuable insight into the industry, including interviews with award-winning cinematographers, a "tricks of the trade" page, information about film schools, multimedia presentations, and the American Cinematographer *online magazine:*
AMERICAN SOCIETY OF CINEMATOGRAPHERS (ASC)
5700 Wilshire Boulevard, Suite 600
Los Angeles, CA 90036
Web: http://www.cinematographer.com

For education and training information, contact:
INTERNATIONAL ALLIANCE OF THEATRICAL STAGE EMPLOYEES, MOVING PICTURE TECHNICIANS, ARTISTS AND ALLIED CRAFTS
1430 Broadway, 20th Floor
New York, NY 10018
Tel: 212-730-1770
Web: http://www.iatse.lm.com

Media Planners
and Buyers

Overview

Media specialists are responsible for placing advertisements that will reach targeted customers and get the best response from the market for the least amount of money. Within the media department, *media planners* gather information about the size and types of audiences that can be reached through each of the various media and about the cost of advertising in each medium. *Media buyers* purchase space in printed publications, as well as time on radio or television stations. Advertising media workers are supervised by a *media director,* who is accountable for the overall media plan. In addition to advertising agencies, media planners and buyers work for large companies that purchase space or broadcast time. There are approximately 155,000 advertising sales agents employed in the United States.

History

The first formal media that allowed advertisers to deliver messages about their products or services to the public were newspapers and magazines, which began selling space to advertisers in the late 19th century. This system of placing ads gave rise to the first media planners and buyers, who were charged with deciding what kind of advertising to put in which publications and then actually purchasing the space.

In the broadcast realm, radio stations started offering program time to advertisers in the early 1900s. And while television advertising began just before the end of World War II, producers were quick to realize that they could reach huge audiences by placing ads on TV. Television advertising proved to be beneficial to the TV stations as well, since they relied on sponsors for financial assistance in order to bring programs into people's homes. In the past, programs were sometimes named not for the host or star of the program, but for the sponsoring company that was paying for the broadcast of that particular show.

During the early years of radio and television, it was often possible for one sponsor to pay for an entire 30-minute program. The cost of producing shows on radio and television, however, increased dramatically, requiring many sponsors to support a single radio or television program. Media planners and buyers learned to get more for their money by buying smaller amounts of time—60, 30, and even 10 seconds—on a greater number of programs.

Today's media planners and buyers have a wide array of media from which to choose. The newest of these, the World Wide Web, allows advertisers not only to precisely target customers but to interact with them as well. In addition to Web banner ads, producers can also advertise via sponsorships, their own Web sites, CD catalogs, voice-mail phone shopping, and more. With so many choices, media planners and buyers must carefully determine target markets and select the ideal media mix in order to reach these markets at the least cost.

The Job

While many employees may work in the media department, the primary specialists are the media planner and the media buyer. They work with professionals from a wide range of media—from billboards, direct mail, and magazines to television, radio, and the Web. Both types of media specialists must be familiar with the markets that each medium reaches, as well as the advantages and disadvantages of advertising in each.

Media planners determine target markets based on their clients' advertising approaches. Considering their clients' products and services, budget, and image, media planners gather information about the public's viewing, reading, and buying habits by administering questionnaires and conducting other forms of market research. Through this research, planners are able to identify target markets by sorting data according to people's ages, incomes, marital status, interests, and leisure activities.

By knowing which groups of people watch certain shows, listen to specific radio stations, or read particular magazines or newspapers, media planners can help clients select air time or print space to reach the consumers most likely to buy their products. For example, Saturday morning television shows attract children, while prime-time programs often draw family audiences. For shows broadcast at these times, media planners will recommend air time to their clients who manufacture products of interest to these viewers, such as toys and automobiles, respectively.

Media planners who work directly for companies selling air time or print space must be sensitive to their clients' budgets and resources. When tailoring their sales pitch to a particular client's needs, planners often go to great lengths to persuade the client to buy air time or advertising space. They produce brochures and reports that detail the characteristics of their viewing or reading market, including the average income of those individuals, the number of people who see the ads, and any other information that may be likely to encourage potential advertisers to promote their products.

Media planners try to land contracts by inviting clients to meetings and presentations and educating them about various marketing strategies. They must not only pursue new clients but also attend to current ones, making sure that they are happy with their existing advertising packages. For both new and existing clients, the media planner's main objective is to sell as much air time or ad space as possible.

Media buyers do the actual purchasing of the time on radio or television or the space in a newspaper or magazine in which an advertisement will run. In addition to tracking the time and space available for purchase, media buyers ensure that ads appear when and where they should, negotiate costs for ad placement, and calculate rates, usage, and budgets. They are also responsible for maintaining contact with clients, keeping them informed of all advertising-related developments and resolving any conflicts that arise. Large companies that generate a lot of advertising or those that place only print ads or only broadcast ads sometimes differentiate between the two main media groups by employing *space buyers* and/or *time buyers.*

Workers who actually sell the print space or air time to advertisers are called *print sales workers* or *broadcast time salespeople.* Like media planners, these professionals are well versed about the target markets served by their organizations and can often provide useful information about editorial content or broadcasted programs.

In contrast to print and broadcast planners and buyers, *interactive media specialists* are responsible for managing all critical aspects of their clients' online advertising campaigns. While interactive media planners may have responsibilities similar to those of print or broadcast planners, they also act as new technology specialists, placing and tracking all online ads and maintaining relationships with clients and Webmasters alike.

The typical online media planning process begins with an agency spreadsheet that details the criteria about the media buy. These criteria often include target demographics, start and end dates for the ad campaign, and online objectives. After sending all relevant information to a variety of Web sites, the media specialist receives cost, market, and other data from the sites. Finally, the media specialist places the order and sends all creative information needed to the selected Web sites. Once the order has been placed, the media specialist receives tracking and performance data and then compiles and analyzes the information in preparation for future ad campaigns.

Media planners and buyers may have a wide variety of clients. Restaurants, hotel chains, beverage companies, food product manufacturers, and automobile dealers all need to advertise to attract potential customers. While huge companies, such as soft drink manufacturers, major airlines, and vacation resorts, pay a lot of money to have their products or services advertised nationally, many smaller firms need to advertise only in their immediate area. Local advertising may come from a health club that wants to announce a special membership rate or from a retail store promoting a sale. Media planners and buyers must be aware of their various clients' advertising needs and create campaigns that will accomplish their promotional objectives.

Requirements

HIGH SCHOOL

Although most media positions, including those at the entry level, require a bachelor's degree, you can prepare for a future job as media planner and/or buyer by taking specific courses offered at the high school level. These include business, marketing, advertising, cinematography, radio and television, and film and video. General liberal arts classes, such as economics, English, communication, and journalism are also important, since media planners and buyers must be able to communicate clearly with both clients and co-workers. In addition, mathematics classes will give you the skills to work accurately with budget figures and placement costs.

POSTSECONDARY TRAINING

Increasingly, media planners and buyers have college degrees, often with majors in marketing or advertising. Even if you have prior work experience or training in media, you should select college classes that provide a good balance of business course work, broadcast and print experience, and liberal arts studies.

Business classes may include economics, marketing, sales, and advertising. In addition, courses that focus on specific media, such as cinematography, film and video, radio and television, and new technologies (like the Internet), are important. Additional classes in journalism, English, and speech will prove helpful as well. Media directors often need to have a master's degree, as well as extensive experience working with the various media.

OTHER REQUIREMENTS

Media planners and buyers in broadcasting should have a keen understanding of programming and consumer buying trends, as well as a knowledge of each potential client's business. Print media specialists must be familiar with the process involved in creating print ads and the markets reached by various publications. In addition, all media workers need to be capable of maintaining good relationships with current clients, as well as pursuing new clients on a continual basis.

Communication and problem solving skills are important, as are creativity, common sense, patience, and persistence. Media planners and buyers must also have excellent oral, written, and analytical skills, knowledge of interactive media planning trends and tools, and the ability to handle multiple assignments in a fast-paced work environment. Strategic thinking skills, industry interest, and computer experience with both database and word processing programs are also vital.

Exploring

Many high schools and two-year colleges and most four-year colleges have media departments that may include radio stations and public access or cable television channels. In order to gain worthwhile experience in media, you can work for these departments as aides, production assistants, programmers, or writers. In addition, school newspapers and yearbooks often need students to sell advertising to local merchants. Theater departments also frequently look for people to sell ads for performance programs.

In the local community, newspapers and other publications often hire high school students to work part-time and/or in the summer in sales and clerical positions for the classified advertising department. Some towns have cable television stations that regularly look for volunteers to operate cameras, sell advertising, and coordinate various programs. In addition, a variety of church- and synagogue-sponsored activities, such as craft fairs, holiday boutiques, and rummage sales, can provide you with opportunities to create and place ads and work with the local media in order to get exposure for the events.

Employers

Media planners and buyers often work for advertising agencies in large cities, such as Chicago, New York, and Los Angeles. These agencies represent various clients who are trying to sell everything from financial services to dishwasher soap. Other media specialists work directly for radio and television networks, newspapers, magazines, and Web sites selling air time and print space. While many of these media organizations are located in large urban areas, particularly radio and television stations, most small towns put out newspapers and therefore need specialists to sell ad space and coordinate accounts.

Starting Out

More than half of the jobs in print and broadcast media do not remain open long enough for companies to advertise available positions in the classified sections of newspapers. As a result, many media organizations, such as radio and television stations, do not usually advertise job openings in the want ads. Media planners and buyers often hear about available positions through friends, acquaintances, or family members and frequently enter the field as entry-level broadcasting or sales associates. Both broadcasting and sales can provide employees just starting out with experience in approaching and working for clients, as well as knowledge about the specifics of programming and its relation to selling air time.

Advertising agencies sometimes do advertise job openings, both in local and national papers and on the Web. Competition is quite fierce for entry-level jobs, however, particularly at large agencies in big cities.

Print media employees often start working on smaller publications as in-house sales staff members, answering telephones and taking orders from customers. Other duties may include handling classified ads or coordinating the production and placement of small print ads created by in-house artists and typesetters. While publications often advertise for entry-level positions, the best way to find work in advertising is to send resumes to as many agencies, publications, and broadcasting offices as possible. With any luck, your resume will arrive just as an opening is becoming available.

While you are enrolled in a college program, you should investigate opportunities for internships or on-campus employment in related areas. Your school's career planning center or placement office should have information on such positions. Previous experience often provides a competitive edge for all job seekers, but it is crucial to aspiring media planners and buyers.

Advancement

Large agencies and networks often hire only experienced people, so it is common for media planners and buyers to learn the business at smaller companies. These opportunities allow media specialists to gain the experience and confidence they need to move up to more advanced positions. Jobs at smaller agencies and television and radio stations also provide possibilities for more rapid promotion than those at larger organizations.

Media planners and buyers climbing the company ladder can advance to the position of media director or may earn promotions to executive-level positions. For those already at the management level, advancement can come in the form of larger clients and more responsibility. In addition, many media planners and buyers who have experience with traditional media are investigating the opportunities and challenges that come with the job of interactive media planner/buyer or Web media specialist.

Earnings

Because media planners and buyers work for a variety of organizations all across the country and abroad, earnings can vary greatly. In general, however, those just entering the field as assistant media planners/traffic coordinators make an average of $17,500 to $22,500 per year, while a media desk coordinator makes $27,500 to $32,500 annually, according to *Advertising Age.*

Media directors can earn between $46,000 and $92,400, depending on the type of employer and the director's experience level. For example, directors at small agencies make an average of $42,100, while those at large agencies can earn up to $100,000, says *Advertising Age.*

Media planners and buyers in television typically earn higher salaries than those in radio. In general, however, beginning broadcasting salespeople usually earn between $18,000 and $35,000 per year and can advance to as much as $46,000 after a few years of experience.

Starting salaries for print advertising salespeople range from $17,500 to $33,000 a year. Experienced workers may earn salaries up to $44,200 a year, on average. Some salespeople draw straight salaries, some receive bonuses that reflect their level of sales, and still others earn their entire wage based on commissions. These commissions are usually calculated as a percentage of sales that the employee brings into the company.

According to the U.S. Bureau of Labor Statistics, advertising sales agents had median annual earnings of $35,850 in 2000. Salaries ranged from $18,570 to $87,240.

Most employers of media planners and buyers offer a variety of benefits, including health and life insurance, a retirement plan, and paid vacation and sick days.

Work Environment

Although media planners and buyers often work a 40-hour week, the hours are not strictly nine to five. Service calls, presentations, and meetings with ad space reps and client are important parts of the job that usually have a profound impact on work schedules. In addition, media planners and buyers must invest considerable time investigating and reading about trends in programming, buying, and advertising.

The variety of opportunities for media planners and buyers results in a wide diversity of working conditions. Larger advertising agencies, publications, and networks may have modern and comfortable working facilities. Smaller markets may have more modest working environments.

Whatever the size of the organization, many planners seldom go into the office and must call in to keep in touch with the home organization. Travel is a big part of media planners' responsibilities to their clients, and they may have clients in many different types of businesses and services, as well as in different areas of the country.

While much of the media planner and buyer's job requires interaction with a variety of people, including co-workers, sales reps, supervisors, and clients, most media specialists also perform many tasks that require independent work, such as researching and writing reports. In any case, the media planner and buyer must be able to handle many tasks at the same time in a fast-paced, continually changing environment.

Outlook

The outlook for employment opportunities for media planners and buyers, like the outlook for the advertising industry itself, depends on the general health of the economy. As the economy thrives, companies produce an increasing number of goods and seek to promote them via newspapers, magazines, television, radio, the Internet, and various other media. The U.S. Department of Labor anticipates that employment in the advertising industry is projected to grow 32 percent over the 2000-10 period, much faster than the average for all occupations.

More and more people are relying on radio and television for their entertainment and information. With cable and local television channels offering a wide variety of programs, advertisers are increasingly turning to TV in order to get exposure for their products and services. Although newspaper sales are in decline, there is growth in special interest periodicals and other print publications. Interactive media, such as the Internet, CD catalogs, and voice-mail shopping, are providing a flurry of advertising activity all around the world. The Web alone promises advertisers exposure in 65 million American households. All of this activity will increase market opportunities for media planners and buyers.

Employment possibilities for media specialists are far greater in large cities, such as New York, Los Angeles, and Chicago, where most magazines and many broadcast networks have their headquarters. However, smaller publications are often located in outlying areas, and large national organizations usually have sales offices in several cities across the country.

Competition for all advertising positions, including entry-level jobs, is expected to be intense. Those employees who have experience, such as media planners and buyers, have the best chances of finding employment.

For More Information

For information on student membership (at the college level), contact:
AMERICAN ADVERTISING FEDERATION (AAF)
1101 Vermont Avenue, NW, Suite 500
Washington, DC 20005-6306
Email: aaf@aaf.org
Web: http://www.aaf.org

The management-oriented national trade organization represents the advertising agency business.
AMERICAN ASSOCIATION OF ADVERTISING AGENCIES
405 Lexington, 18th Floor
New York, NY 10174-1801
Web: http://www.aaaa.org

Radio and Television Announcers

Quick Facts

School Subjects
English
Speech
Personal Skills
Communication/ideas
Following instructions
Work Environment
Primarily indoors
Primarily one location
Minimum Education Level
Some postsecondary training
Salary Range
$10,000 to $26,000 to $200,000
Certification or Licensing
None available
Outlook
Decline

Overview

Radio and television announcers present news and commercial messages from a script. They identify the station, announce breaks, and introduce and wrap up shows. Interviewing guests and giving public service announcements may also be part of the work. At small stations, the announcer may keep the program log, run the transmitter, and cue the changeover to network broadcasting as well as write scripts or rewrite news releases. About 71,000 people are employed as radio and television announcers and newscasters in the United States.

History

Guglielmo Marconi, a young Italian engineer, first transmitted a radio signal in his home in 1895. Radio developed rapidly as people began to comprehend the tremendous possibilities. The stations KDKA in Pittsburgh and WWJ in Detroit began broadcasting in 1920. Within 10 years, there were radio stations in all the major cities in the United States and broadcasting had become big business. The National Broadcasting Company became the first network in 1926 when it linked together 25 stations across the country. The Columbia Broadcasting System was organized in the following year. In 1934, the Mutual Broadcasting Company was founded. The years between 1930 and 1950 may be considered the zenith years of the radio industry. With the coming of television, radio broadcasting took second place in importance as entertainment for the home—but radio's commercial and communications value should not be underestimated.

Discoveries that led to the development of television can be traced as far back as 1878, when William Crookes invented a tube that produced the cathode ray. Other inventors who contributed to the development of television were Vladimir Zworykin, a Russian-born scientist who came to this country at the age of 20 and is credited with inventing the iconoscope before he was 30; Charles Jenkins, who invented a scanning disk, using certain vacuum tubes and photoelectric cells; and Philo Farnsworth, who invented an image dissector. WNBT and WCBW, the first commercially licensed television stations, went on the air in 1941 in New York. Both suspended operations during World War II, but resumed them in 1946 when television sets began to be manufactured on a commercial scale.

As radio broadcasting was growing across the country in its early days, the need for announcers grew. They identified the station and brought continuity to broadcast time by linking one program with the next as well as participating in many programs. In the early days (and even today in smaller stations), announcers performed a variety of jobs around the station. When television began, many radio announcers and newscasters started to work in the new medium. The need for men and women in radio and television broadcasting has continued to grow. Television news broadcasting requires specialized "on-camera" personnel-anchors, television news reporters, broadcast news analysts, consumer reporters, and sports reporters (sportscasters).

The Job

Some announcers merely announce; others do a multitude of other jobs, depending on the size of the station. But the nature of their announcing work remains the same.

An announcer is engaged in an exacting career. The necessity for finishing a sentence or a program at a precisely planned moment makes this a demanding and often tense career. It is absolutely essential that announcers project a sense of calm to their audiences, regardless of the activity and tension behind the scenes.

The announcer who plays recorded music interspersed with a variety of advertising material and informal commentary is called a disc jockey. This title arose when most music was recorded on conventional flat records or discs. Today much of the recorded music used in commercial radio stations is on magnetic tape or compact discs. *Disc jockeys* serve as a bridge between the music itself and the listener. They may perform such public services as announcing the time, the weather forecast, or important news. It can be a lonely job, since many disc jockeys are the only person in the studio. But because their job is to maintain the good spirits of their audience and to attract new listeners, disc jockeys must possess the ability to be relaxed and cheerful. (For more information on this career, see the article, Disc Jockeys.)

Unlike the more conventional radio or television announcer, the disc jockey is not bound by a written script. Except for the commercial announcements, which must be read as written, the disc jockey's statements are usually spontaneous. Disc jockeys usually are not required to play a musical selection to the end; they may fade out a record when it interferes with a predetermined schedule for commercials, news, time checks, or weather reports.

Announcers who cover sports events for the benefit of the listening or viewing audience are known as *sportscasters,* or *sports broadcasters and announcers.* This is a highly specialized form of announcing as sportscasters must have extensive knowledge of the sports that they are covering, plus the ability to describe events quickly and accurately.

Often the sportscaster will spend several days with team members, observing practice sessions, interviewing people, and researching the history of an event or of the teams to be covered. The more information that a sportscaster can acquire about individual team members, company they represent, tradition of the contest, ratings of the team, and community in which the event takes place, the more interesting the coverage is to the audience. (For more information on this career, see the article Sports Broadcasters and Announcers.)

The announcer who specializes in reporting the news to the listening or viewing public is called a *newscaster.* This job may require simply reporting facts, or it may include editorial commentary. Newscasters may be given the authority by their employers to express their opinions on news items or the philosophies of others. They must make judgments about which news is important and which is not. In some instances, they write their own scripts, based on facts that are furnished by international news bureaus. In other instances, they read text exactly as it comes in over a teletype machine. They may make as few as one or two reports each day if they work on a major news program, or they may broadcast news for five minutes every hour or half-hour. Their delivery is usually dignified, measured, and impersonal.

The *anchor* generally summarizes and comments on one aspect of the news at the end of the scheduled broadcast. This kind of announcing differs noticeably from that practiced by the sportscaster, whose manner may be breezy and interspersed with slang, or from the disc jockey, who may project a humorous, casual, or intimate image.

The newscaster may specialize in certain aspects of the news, such as economics, politics, or military activity. Newscasters also introduce films and interviews prepared by *news reporters* (see the article, Reporters) that provide in-depth coverage and information on the event being reported. *News analysts* are often called *commentators,* and they interpret specific events and discuss how these may affect individuals or the nation. They may have a specified daily slot for which material must be written, recorded, or presented live. They gather information that is analyzed and interpreted through research and interviews and cover public functions such as political conventions, press conferences, and social events.

Smaller television stations may have an announcer who performs all the functions of reporting, presenting, and commenting on the news as well as introducing network and news service reports.

Many television and radio announcers have become well-known public personalities in broadcasting. They may participate in community activities as master of ceremonies at banquets and other public events.

Requirements

HIGH SCHOOL
Although there are no formal educational requirements for entering the field of radio and television announcing, many large sta-

tions prefer college-educated applicants. The general reason given for this preference is that announcers with broad educational and cultural backgrounds are better prepared to successfully meet a variety of unexpected or emergency situations. The greater the knowledge of geography, history, literature, the arts, political science, music, science, and of the sound and structure of the English language, the greater the announcer's value.

In high school, therefore, you should focus on a college preparatory curriculum, according to Steve Bell, a professor of telecommunications at Ball State University. A former network anchor who now teaches broadcast journalism, he says, "One trend that concerns me is that some high schools are developing elaborate radio and television journalism programs that take up large chunks of academic time, and I think that is getting the cart before the horse. There's nothing wrong with one broadcast journalism course or extracurricular activities, but not at the expense of academic hours."

In that college preparatory curriculum, you should learn how to write and use the English language in literature and communication classes. Subjects such as history, government, economics, and a foreign language are also important.

POSTSECONDARY TRAINING

When it comes to college, having your focus in the right place is essential, according to Professor Bell. "You want to be sure you're going to a college or university that has a strong program in broadcast journalism, where they also put a strong emphasis on the liberal arts core."

Some advocate a more vocational type of training in preparation for broadcast journalism, but Bell cautions against strictly vocational training. "The ultimate purpose of college is to have more of an education than you have from a trade school. It is important to obtain a broad-based understanding of the world we live in, especially if your career goal is to become an anchor."

A strong liberal arts background with emphasis in journalism, English, political science, or economics is advised, as well as a telecommunications or communications major.

OTHER REQUIREMENTS

A pleasing voice and personality are of great importance to prospective announcers. They must be levelheaded and able to react calmly in the face of a major crisis. People's lives may depend on an announcer's ability to remain calm during a disaster. There are also many unexpected circumstances that demand the skill of quick thinking. For example, if guests who are to appear on a program do not arrive or become too nervous to go on the air, the announcer must compensate immediately and fill the airtime. He or she must smooth over an awkward phrase, breakdown in equipment, or other technical difficulty.

Good diction and English usage, thorough knowledge of correct pronunciation, and freedom from regional dialects are very important. A factual error, grammatical error, or mispronounced word can bring letters of criticism to station managers.

Those who aspire to careers as television announcers must present a good appearance and have no nervous mannerisms. Neatness, cleanliness, and careful attention to the details of proper dress are important. The successful television announcer must have the combination of sincerity and showmanship that attracts and captures an audience.

Broadcast announcing is a highly competitive field. Although there may not be any specific training program required by prospective employers, station officials pay particular attention to taped auditions of an applicant's delivery or, in the case of television, to videotaped demos of sample presentations.

A Federal Communications Commission license or permit is no longer required for broadcasting positions. Union membership may be required for employment with large stations in major cities and is a necessity with the networks. The largest talent union is the American Federation of Television and Radio Artists (AFTRA). Most small stations, however, are nonunion.

Exploring

If a career as an announcer sounds interesting, try to get a summer job at a radio or television station. Although you will probably not have the opportunity to broadcast, you may be able to judge whether or not the type of work appeals to you as a career.

Any chance to speak or perform before an audience should be welcomed. Appearing as a speaker or performer can show whether or not you have the stage presence necessary for a career in front of a microphone or camera.

Many colleges and universities have their own radio and television stations and offer courses in radio and television. You can gain valuable experience working at college-owned stations. Some radio stations, cable systems, and TV stations offer financial assistance, internships, and co-op work programs, as well as scholarships and fellowships.

Employers

Of the roughly 71,000 radio and television announcers working in the United States, almost all are on staff at one of the 13,012 radio stations or 1,686 television stations around the country. Some, however, work on a freelance basis on individual assignments for networks, stations, advertising agencies, and other producers of commercials. Worldwide, as of January 2000, there were an estimated 21,500 television stations and more than 44,000 radio stations.

Some companies own several television or radio stations; some belong to networks such as ABC, CBS, NBC, or FOX, while others are independent. While radio and television stations are located throughout the United States, major markets where better paying jobs are found, are generally near large metropolitan areas.

Starting Out

One way to enter this field is to apply for an entry-level job rather than an announcer position. It is also advisable to start at a small station. Most announcers start in jobs such as production secretary, production assistant, researcher, or reporter in small stations. As opportunities arise, they move from one job to another. Work as a disc jockey, sportscaster, or news reporter may become available. Network jobs are few, and the competition for them is great. An announcer must have several years of experience as well as a college education to be considered for these positions.

An announcer is employed only after an audition. Applicants should carefully select audition material to show a prospective employer the full range of one's abilities. In addition to presenting prepared materials, applicants may be asked to read material that they have not seen previously, such as a commercial, news release, dramatic selection, or poem.

Advancement

Most successful announcers advance from small stations to large ones. Experienced announcers usually have held several jobs. The most successful announcers may be those who work for the networks. Usually, because of network locations, announcers must live in or near the country's largest cities.

Some careers lead from announcing to other aspects of radio or television work. More people are employed in sales, promotion, and planning than in performing; often they are paid more than announcers. Because the networks employ relatively few announcers in proportion to the rest of the broadcasting professionals, a candidate must have several years of experience and specific background in several news areas before being considered for an audition. These top announcers generally are college graduates.

Earnings

According to a 2001 salary survey by the Radio-Television News Directors Association, there is a wide range of salaries for announcers. For radio announcers, the median salary was $25,000 with a low of $10,000 and a high of $41,500. For television reporters and announcers, the median salary was $26,000 with a low of $14,000 and a high of $200,000.

The U.S. Bureau of Labor Statistics reports that median annual earnings of announcers in 2000 were $19,800. Salaries ranged from $12,360 to $50,660.

For both radio and television, salaries are higher in the larger markets. Salaries are also generally higher in commercial than in public broadcasting. Nationally known announcers and newscasters who appear regularly on network television programs receive salaries that may be quite impressive. For those who become top television personalities in large metropolitan areas, salaries also are quite rewarding.

Most radio or television stations broadcast 24 hours a day. Although much of the material may be prerecorded, announcing staff must often be available and as a result, may work considerable overtime or split shifts, especially in smaller stations. Evening, night, weekend, and holiday duty may provide additional compensation.

Work Environment

Work in radio and television stations is usually very pleasant. Almost all stations are housed in modern facilities. The maintenance of technical electronic equipment requires temperature and dust control, and people who work around such equipment benefit from the precautions taken to preserve it.

Announcers' jobs may provide opportunities to meet well-known or celebrity persons. Being at the center of an important communications medium can make the broadcaster more keenly aware of current issues and divergent points of view than the average person.

Announcers and newscasters usually work a 40-hour week, but they may work irregular hours. They may report for work at a very early hour in the morning or work late into the night. Some radio stations operate on a 24-hour basis. All-night announcers may be alone in the station during their working hours.

Outlook

Competition for entry-level employment in announcing during the coming years is expected to be keen as the broadcasting industry always attracts more applicants than are needed to fill available openings. There is a better chance of working in radio than in television because there are more radio stations. Local television stations usually carry a high percentage of network programs and need only a very small staff to carry out local operations.

The U.S. Department of Labor predicts that opportunities for experienced broadcasting personnel will decline through 2010 due to the slowing growth of new radio and television stations. Openings will result mainly from those who leave the industry or the labor force. The trend among major networks, and to some extent among many smaller radio and TV stations, is toward specialization in such fields as sportscasting or weather forecasting. Newscasters who specialize in such areas as business, consumer, and health news should have an advantage over other job applicants.

For More Information

For information on its summer internship program, contact:
ASSOCIATION OF LOCAL TELEVISION STATIONS
1320 19th Street, NW, Suite 300
Washington, DC 20036
Tel: 202-887-1970
Web: http://www.altv.com

For a list of schools offering degrees in broadcasting as well as scholarship information, contact:
BROADCAST EDUCATION ASSOCIATION
1771 N Street, NW
Washington, DC 20036-2891
Tel: 888-380-7222
Email: beainfo@beaweb.org
Web: http://www.beaweb.org

For college programs and union information, contact:
NATIONAL ASSOCIATION OF BROADCAST EMPLOYEES AND TECHNICIANS
501 Third Street, NW, 8th Floor
Washington, DC 20001
Tel: 202-434-1254
Email: nabet@nabetcwa.org
Web: http://nabetcwa.org

For broadcast education, support, and scholarship information, contact:
NATIONAL ASSOCIATION OF BROADCASTERS
1771 N Street, NW
Washington, DC 20036
Tel: 202-429-5300
Email: nab@nab.org
Web: http://www.nab.org

For general information, contact:
NATIONAL ASSOCIATION OF FARM BROADCASTERS
26 East Exchange Street, Suite 107
St. Paul, MN 55101
Tel: 651-224-0508
Email: info@nafb.com
Web: http://nafb.com

For a booklet on careers in cable, contact:
NATIONAL CABLE TELEVISION ASSOCIATION
1724 Massachusetts Avenue, NW
Washington, DC 20036
Tel: 202-775-3550
Web: http://www.ncta.com

For scholarship and internship information, contact:
RADIO-TELEVISION NEWS DIRECTORS ASSOCIATION
RADIO-TELEVISION NEWS DIRECTORS FOUNDATION
1600 K Street, NW, Suite 700
Washington, DC 20006-2838
Tel: 202-659-6510
Email: rtnda@rtnda.org or rtndf@rtndf.org
Web: http://www.rtnda.org

Radio and Television Program Directors

Overview

Radio and television rogram direc-tors plan and schedule program material for radio and television stations and networks. They determine the entertainment programs, news broadcasts, and other program material their organizations offer to the public. At a large network the program director may supervise a large programming staff. At a small station one person may manage the station and also handle all programming duties.

History

Radio broadcasting in the United States began after World War I. The first commercial radio station, KDKA in Pittsburgh, came on the air in 1920 with a broadcast of presidential election returns. About a dozen radio stations were broadcasting by 1921. In 1926, the first national network linked stations across the country. Today there are over 13,012 commercial and public radio stations in the United States.

The first public demonstration of television in the United States came in 1939 at the opening of the New York World's Fair. Further development was limited during World War II, but by 1953 there were about 120 stations. By the end of 2001, the United States had 1,686 commercial, public, and cable television stations. According to the National Cable Television Association, between

1996 and 2001, the number of cable networks increased by 91 percent, from 147 to 281 channels.

The Job

Program directors plan and schedule program material for radio and television stations and networks. They work in both commercial and public broadcasting and may be employed by individual radio or television stations, regional or national networks, or cable television systems.

The material program directors work with includes entertainment programs, public service programs, newscasts, sportscasts, and commercial announcements. Program directors decide what material is broadcast and when it is scheduled; they work with other staff members to develop programs and buy programs from independent producers. They are guided by such factors as the budget available for program material, the audience their station or network seeks to attract, their organization's policies on content and other matters, and the kinds of products advertised in the various commercial announcements.

In addition, program directors may set up schedules for the program staff, audition and hire announcers and other on-the-air personnel, and assist the sales department in negotiating contracts with sponsors of commercial announcements. The duties of individual program directors are determined by such factors as whether they work in radio or television, for a small or large organization, for one station or a network, or in a commercial or public operation.

At small radio stations the owner or manager may be responsible for programming, but at larger radio stations and at television stations the staff usually includes a program director. At medium to large radio and television stations the program director usually has a staff that includes such personnel as music librarians, music directors, editors for tape or film segments, and writers. Some stations and networks employ *public service directors*. It is the responsibility of these individuals to plan and schedule radio or television public service programs and announcements in such fields as education, religion, and civic and government affairs. Networks often employ *broadcast operations directors,* who coordinate the activities of the personnel who prepare network program schedules, review program schedules, issue daily corrections, and advise affiliated stations on their schedules.

Program directors must carefully coordinate the various elements for a station while keeping in tune with the listeners, viewers, advertisers, and sponsors.

Other managers in radio and television broadcasting include production managers, operations directors, news directors, and sports directors. The work of program directors usually does not include the duties of *radio directors* or *television directors,* who direct rehearsals and integrate all the elements of a performance.

Requirements

HIGH SCHOOL
If you are interested in this career, you should take courses that develop your communication skills in high school. Such classes include English, debate, and speech. You also should take business courses to develop your management skills; current events and history courses to develop your understanding of the news and the trends that affect the public's interests; and such courses as dance, drama, music, and painting to expand your understanding of the creative arts. Finally, don't neglect your computer skills. You will probably be using computers throughout your career to file reports, maintain schedules, and plan future programming projects.

POSTSECONDARY TRAINING
Those with the most thorough educational backgrounds will find it easiest to advance in this field. A college degree, therefore, is recommended for this field. Possible majors for those interested in this work include radio and television production and broadcasting, communications, liberal arts, or business administration. You will probably take English, economics, business administration, computer, and media classes. You may also wish to acquire some technical training that will help you understand the engineering aspects of broadcasting.

OTHER REQUIREMENTS
Program directors must be creative, alert, and adaptable people who stay up-to-date on the public's interests and attitudes and are

able to recognize the potential in new ideas. They must be able to work under pressure and be willing to work long hours, and they must be able to work with all kinds of people. Program directors also must be good managers who can make decisions, oversee costs and deadlines, and attend to details.

Exploring

If your high school or college has a radio or television station, you should volunteer to work on the staff. You also should look for part-time or summer jobs at local radio or television stations. You may not be able to plan the programming at a local station, but you will see how a station works and be able to make contacts with those in the field. If you can't find a job at a local station, at least arrange for a visit and ask to talk to the personnel. You may be able to "shadow" a program director for a day—that is, follow that director for the work day and see what his or her job entails.

Employers

According to the National Association of Broadcasters, there were 1,686 broadcast television stations and 13,012 radio stations in the United States in 2001. Cable television stations add another option for employment.

Large conglomerates own some stations while others are owned individually. While radio and television stations are located all over the country, the largest stations with the highest paid positions are located in large metropolitan areas.

Starting Out

Program director jobs are not entry-level positions. A degree and extensive experience in the field is required. Most program directors have technical and on-air experience in either radio or television. While you are in college, you should investigate the availability of internships since internships are almost essential for prospective job candidates. Your college placement office should also have information on job openings. Private and state employment agencies may also prove useful resources. You can also send resumes to radio and television stations or apply in person.

Beginners should be willing to relocate as they are unlikely to find employment in large cities. They usually start at small stations with fewer employees, allowing them a chance to learn a variety of skills.

Advancement

Most beginners start in entry-level jobs and work several years before they have enough experience to become program directors. Experienced program directors usually advance by moving from small stations to larger stations and networks or by becoming station managers.

Earnings

Salaries for radio and television program directors vary widely based on such factors as size and location of the station, whether the station is commercial or public, and experience of the director. Yearly earnings for radio program directors may range from approximately $25,000 to $100,000, while those for television program directors generally range from approximately $25,000 to $125,000. According to the U.S. Bureau of Labor Statistics, median annual earnings of producers and directors in radio and television broadcasting were $34,630 in 2000. Salaries for all producers and directors ranged from $21,050 to $87,770. According to the 2001 Radio and Television Salary Survey by Radio-Television News Directors Association, radio news directors earned a median of $31,000, and salaries ranged from a low of $10,000 to a high of $125,000. Television news directors earned a median of $65,000, with salaries ranging from $21,000 to $250,000. Assistant news directors earned between $30,000 and $125,000, with a median of $54,500.

Television stations usually pay higher salaries than radio stations, and a program director for a large network station may make hundreds of thousands of dollars per year. Both radio and television program directors usually receive health and life coverage benefits and sometimes receive yearly bonuses as well.

Work Environment

Program directors at small stations often work 44 to 48 hours a week and frequently work evenings, late at night, and weekends. At larger stations, which have more personnel, program directors usually work 40-hour weeks.

Program directors frequently work under pressure because of the need to maintain precise timing and meet the needs of sponsors, performers, and other staff members.

Although the work is sometimes stressful and demanding, program directors usually work in pleasant environments with creative staffs. They also interact with the community to arrange programming and deal with a variety of people.

Outlook

Today more than 13,000 radio and television stations, cable television systems, and regional and national networks employ program directors or have other employees whose duties include programming. According to the U.S. Department of Labor, employment in radio and television broadcasting is expected to increase only 10 percent over the 2000-10 period, slower than the average for all industries. Slow growth rate is attributed to industry consolidation, introduction of new technologies, greater use of prepared programming, and competition from other media.

Competition for radio and television program director jobs is strong. There are more opportunities for beginners in radio than there are in television. Most radio and television stations in large cities only hire experienced workers.

New radio and television stations and new cable television systems are expected to create additional openings for program directors, but some radio stations are eliminating program director positions by installing automatic programming equipment or combining those responsibilities with other positions.

For More Information

This organization has industry news for those involved in local, over-the-air television broadcasting.
ASSOCIATION OF LOCAL TELEVISION STATIONS
1320 19th Street, NW, Suite 300
Washington, DC 20036
Tel: 202-887-1970
Email: altv@aol.com
Web: http://www.altv.com

For a list of schools offering degrees in broadcasting, contact:
BROADCAST EDUCATION ASSOCIATION
1771 N Street, NW
Washington, DC 20036-2891
Tel: 202-429-5354
Web: http://www.beaweb.org

For college programs and union information, contact:
NATIONAL ASSOCIATION OF BROADCAST EMPLOYEES AND TECHNICIANS
501 3rd Street, NW, 8th Floor
Washington, DC 20001
Tel: 800-882-9174
Email: nabet@nabetcwa.org
Web: http://nabetcwa.org

For broadcast education, support, and scholarship information, contact:
NATIONAL ASSOCIATION OF BROADCASTERS
1771 N Street, NW
Washington, DC 20036
Tel: 202-429-5300
Email: nab@nab.org
Web: http://www.nab.org

For information on student membership, contact:
NATIONAL ASSOCIATION OF FARM BROADCASTERS
26 Exchange Street East, Suite 307
St. Paul, MN 55101
Tel: 651-224-0508
Email: nafboffice@aol.com
Web: http://nafb.com

For a booklet on careers in cable, contact:
NATIONAL CABLE TELEVISION ASSOCIATION
1724 Massachusetts Avenue, NW
Washington, DC 20036
Tel: 202-775-3550
Web: http://www.ncta.com

For scholarship and internship information, contact:
RADIO-TELEVISION NEWS DIRECTORS ASSOCIATION
RADIO-TELEVISION NEWS DIRECTORS FOUNDATION
1600 K Street, NW, Suite 700
Washington, DC 20006-2838
Tel: 202-659-6510
Email: rtnda@rtnda.org
Web: http://www.rtnda.org

Radio Producers

Overview

Radio producers plan, rehearse, and produce live or recorded programs. They work with the music, on-air personalities, sound effects, and technology to put together an entire radio show. They schedule interviews and arrange for promotional events.

According to the Federal Communications Commission, the United States alone has 13,012 radio stations. Larger stations employ radio producers while smaller stations may combine those duties with those of the program director or disc jockey. While most radio producers work at radio stations, some work to produce a particular show and then sell that show to various stations.

History

As long as radio has existed people have been behind the scenes to make sure that what the audience hears is what the station wants them to hear. A wide variety of administrative, programming, and technical people work behind the scenes of radio shows to create a professional broadcast.

Scheduled broadcasting began with a program broadcast by radio station KDKA in Pittsburgh, and by 1923, 2.5 million radios had been purchased. In the 1930s, radio personalities were household names, and even then, numerous people worked behind the scenes, arranging interviews and coordinating production.

Before television, radio producers would direct the on-air soap operas as well as the news, weather, and music. With the added technology of today's radio broadcast, radio producers are even more important in mixing the special effects, locations, personalities, and formats in a way that creates a good radio show.

The Internet has made the radio producer's job easier in some ways and more challenging in others. Web sites specifically for producers provide a community where ideas can be exchanged for shows, news, jokes, and more. However, with the new frontier of broadcasting on the Internet, radio producers have one more duty to add to their long list of responsibilities.

The Job

The identity and style of a radio program is a result of the collaborations of on-air and off-air professionals. Radio disc jockeys talk the talk during a broadcast, and producers walk the walk behind the scenes. But in many situations, particularly with smaller radio stations, the disc jockey and the show's producer are the same person.

Also, many show producers have disc jockey experience. This experience, combined with technical expertise, helps producers effectively plan their shows.

Brent Lee, a radio producer for WFMS, a country radio station in Indianapolis, began his career while still in high school at the small radio station in his hometown. This early on-air experience, combined with his degree in telecommunications and political science from Ball State University, helped to give Lee the necessary background for his current position.

Radio producers rely on the public's very particular tastes—differences in taste allow for many different kinds of radio to exist, to serve many different segments of a community. In developing radio programs, producers take into consideration the market-place—they listen to other area radio stations and determine what's needed and appreciated in the community, and what there may already be too much of. They conduct surveys and interviews to find out what the public wants to hear. They decide which age groups they want to pursue, and develop a format based on what appeals to these listeners. This all results in a station's identity, which is very important. Listeners associate a station with the kind of music it plays, how much music it plays, the type of news and conversation presented, and the station's on-air personalities.

Based on this feedback, and on market research, radio disc jockeys/producers devise music playlists and music libraries. They each develop an individual on-air identity, or personality. And they invite guests who will interest their listeners. Keeping a show running on time is also the responsibility of a producer. This involves carefully weaving many different elements into a show, including music, news reports, traffic reports, and interviews.

As the producer of the "Jim, Kevin, and Bill Show," Lee arrives at the station at about 4:15 AM each morning to prep for the morning show. The show broadcasts from 5 AM to 9 AM each morning and considers its main audience to be the "morning drivers" on their way to work or school.

The time of the broadcast is one key to planning a radio show. Because of the typical listeners of the morning show, traffic reports are given every 10 minutes. These reports are mixed with weather, news, and music. While the rest of the day, WFMS listeners will hear 13 songs each hour, the morning show typically plays between six and eight songs per hour. Those songs are interspersed with six traffic reports, four weather forecasts, a variety of national, local, and entertainment news, and the typical morning disc jockey banter.

The audience of the country music radio station is mostly female, and the morning show is billed as "good, clean fun" by the station, promoting the family nature of the program.

In addition to keeping in touch with the listening public, producers also keep track of current events. They consult newspapers and other radio programs to determine what subjects to discuss on their morning shows. One of the newest tools that Lee uses is a Web site designed specifically for morning shows. The site provides a forum to share ideas and ask questions.

"There are a couple of things each day that can be used," says Lee. "Since we're a family show, we have to throw some of it out, but it's a really good resource."

Radio producers write copy for and coordinate on-air commercials, which are usually recorded in advance. They also devise contests, from large public events to small, on-air trivia competitions.

Though a majority of radio stations have music formats, radio producers also work for 24-hour news stations, public broadcasting, and talk radio. Producing news programs and radio documentaries involves a great deal of research, booking guests, writing scripts, and interviewing.

"One of the most attractive qualities about this job is it's fun," says Lee. "Each day, I spend half of my first five hours laughing."

Requirements

HIGH SCHOOL
Writing skills are valuable in any profession, but especially in radio. Take composition and literature courses, and other courses that require essays and term papers. Journalism courses will not only help you develop your writing skills, but will teach you about the nature and history of media. You'll learn about deadlines and how to put a complete project (such as a newspaper or yearbook) together.

If your school has a radio station, get involved with it in any way you can. Check with your local radio stations; some may offer part-time jobs to high school students interested in becoming producers and disc jockeys.

Business courses and clubs frequently require students to put together projects; starting any business is similar to producing your own radio show. Use such a project as an opportunity to become familiar with the market research, interviewing, and writing that are all part of a radio producer's job. For both the future radio producer and the future disc jockey, a theater department offers great learning opportunities. Drama or theater classes, which are frequently involved in productions, may provide opportunities for learning about funding, advertising, casting, and other fundamentals similar to a radio production.

POSTSECONDARY TRAINING
Most journalism and communications schools at universities offer programs in broadcasting. Radio producers and announcers often start their training in journalism schools, and they receive hands-on instruction at campus radio stations. These broadcasting programs are generally news-centered, providing great opportunities for students interested in producing news programs, daily newscasts, and documentaries. News directors and program managers of radio stations generally want to hire people who have a good,

well-rounded education with a grounding of history, geography, political science, and literature.

OTHER REQUIREMENTS
Radio producers should be well versed in the English language (or the language they broadcast in), and be creative thinkers who can combine several elements into one project. The ability to understand technical equipment and coordinate it with on-air events is necessary.

A healthy curiosity about people and the world will help radio producers find new topics for news shows, new guests for call-ins, and new ideas for music formats. There are no physical requirements to be a radio producer, although those starting as disc jockeys need a strong, clear voice to be heard over the airwaves.

Exploring

Getting your feet wet early is a good possibility for all radio careers. Small radio stations are often willing to let young, inexperienced people work either behind-the-scenes or on-air. Getting a job or an internship at one of the small stations in your area may be as simple as asking for one.

Many high schools and universities have on-site radio stations where students can get hands-on experience at all different levels. As you explore the career further, you might want to talk with a radio producer about his or her job to see if you are still interested.

Since most people don't start out as a producer, experience in any area of radio is helpful, so talk to local disc jockeys or program directors as well.

Employers

There has been a steady growth in the number of radio stations in the United States. According to the Federal Communications Commission (FCC), there were 13,012 radio stations as of March 2001 and over 3,800 station owners.

However, many stations combine the position of radio producer with that of the disc jockey or program director, so depending on the size of the station and market, producers may or may not be able to find a suitable employer.

Due to the Telecommunications Act of 1996, companies can own an unlimited number of radio stations nationwide with an eight-station limit within one market area, depending on the size of the market. When this legislation took effect, mergers and acquisitions changed the face of the radio industry. So, while the pool of employers is smaller, the number of stations continues to rise.

Starting Out

Radio producers usually start work at radio stations in any capacity possible. After working for a while in a part-time position gaining experience and making connections, a young, dedicated producer will find opportunities to work in production or on-air.

Both experience and a college education are generally needed to become a radio producer. It is best if both the experience and the education are well rounded with exposure to on-air and off-air positions as well as a good working knowledge of the world in which we live.

Although some future producers begin their first radio jobs in paid positions, many serve unpaid internships or volunteer to help run their college or high school station. Even if this entry-level work is unpaid, the experience gained is one of the key necessities to furthering a career in any type of radio work.

With experience as a disc jockey or behind-the-scenes person, an aspiring radio producer might try to land a position at another station, but more likely within a station and format they are used to.

Advancement

Radio producers are a key link in putting together a radio show. Once they have the experience coordinating all the elements that go into a radio production, it is possible to move into a program director (see the article, Radio and Television Program Directors) position or, possibly in the future, to general manager.

Another way to advance is to move from being the producer of a small show to a larger one, or move from a small station to a larger one. Some producers move into the freelance arena, producing their own shows that they sell to several radio stations.

Earnings

According to the Radio-Television News Directors Association 2001 Salary Survey, radio news producers reported salaries ranging from $15,000 to $40,000 per year, with a median of $26,000. Like many radio jobs, there is a wide range resulting from differences in market size and station size of each radio station. Salaries for radio producers are relatively flat, according to the survey, with little growth over the previous year.

Most large stations offer employees typical fringe benefits, although part-time employees may not be eligible for those.

Work Environment

Radio producers generally work indoors in a busy environment, although some location and outdoor work might be required. The atmosphere at a radio station is generally very pleasant; however, smaller stations may not be modern with much of the investment going into high-tech equipment for the broadcasts.

Full-time radio producers usually work more than 40 hours per week planning, scheduling, and producing radio shows. Also, according to the schedule of their shows, early morning, late night, or weekend work might be required. Radio is a 24-hour-a-day, seven-day-a-week production, requiring constant staffing.

Producers work with disc jockeys and program directors in planning radio shows, and they also work with advertising personnel to produce radio commercials. In addition to this collaboration, they may work alone doing research for the show. Working with the public is another aspect of the radio producer's job. Promotions and events may require contact with the people in the business and listeners.

Outlook

The U.S. Department of Labor predicts slower-than-average employment growth in the radio industry through 2010. In the past, radio station ownership was highly regulated by the government, limiting the number of stations a person or company could own. Recent deregulation has made multiple station ownership possible.

Radio stations now are bought and sold at a more rapid pace. This may result in a radio station changing formats, as well as entire staffs. Though some radio producers are able to stay at a station over a period of several years, people going into radio should be prepared to change employers at some point in their careers.

Another trend that is affecting radio producing jobs is the increasing use of programming created by services outside the broadcasting industry. Satellite radio, in which subscribers pay a monthly fee for access to 100 radio stations, will be a big threat to smaller, marginal stations.

Competition is stiff for all radio jobs. Graduates of college broadcasting programs are finding a scarcity of work in media. Paid internships will also be difficult to find—many students of radio will have to work for free for a while to gain experience. Radio producers may find more opportunities as freelancers, developing their own programs independently and selling them to stations.

For More Information

For a list of schools offering degrees in broadcasting as well as scholarship information, contact:
BROADCAST EDUCATION ASSOCIATION
1771 N Street, NW
Washington, DC 20036-2891
Tel: 202-429-5354
Web: http://www.beaweb.org

For college programs and union information, contact:
NATIONAL ASSOCIATION OF BROADCAST EMPLOYEES AND TECHNICIANS
501 3rd Street, NW, 8th Floor
Washington, DC 20001
Tel: 800-882-9174
Email: nabet@nabetcwa.org
Web: http://nabetcwa.org

For broadcast education, support, and scholarship information, contact:

NATIONAL ASSOCIATION OF BROADCASTERS
1771 N Street, NW
Washington, DC 20036
Tel: 202-429-5300
Email: nab@nab.org
Web: http://www.nab.org

For scholarship and internship information, contact:

RADIO-TELEVISION NEWS DIRECTORS ASSOCIATION
RADIO-TELEVISION NEWS DIRECTORS FOUNDATION
1600 K Street, NW, Suite 700
Washington, DC 20006-2838
Tel: 202-659-6510
Email: rtnda@rtnda.org
Web: http://www.rtnda.org

Real-Time Captioners

Overview

Real-time captioners operate a computer-aided transcription (CAT) stenotype system to create closed captions for use in live television broadcasts, in classroom instruction, or in other scenarios requiring live translating or interpreting on the computer. Computer-Aided Real-time Translation, or CART, refers to use of machine steno shorthand skills to produce real-time text on a computer. Generally, captioning systems use a modified stenotype machine connected to a computer. The real-time captioner inputs the captions phonetically (transcription or speech sounds) on the steno machine, and the sounds are then translated into English words by the computer using a special dictionary created by the captioner. During a live broadcast, the captions are entered as the program progresses, much as a court reporter transcribes a trial as it progresses. The input data is sent along telephone lines to the broadcast point, where the caption codes become part of the television signal.

History

Real-time captioning technology arose from a need to make live broadcasts accessible to deaf and hard-of-hearing people. To meet this need, the National Captioning Institute (NCI), founded in 1979, became the chief architect of the computer-based technology needed to bring captions to real-time audiences nationwide. At first, NCI, headquartered in Vienna, Virginia, provided captions only for prerecorded programs. Captions were prepared in advance by people who were not court reporters. It soon became apparent, however, that captions were needed for live television, so NCI went to work developing a system that could prepare captions for live broadcast.

NCI first introduced real-time captioning to eager audiences in April 1982 when it captioned the Academy Awards. Today, real-time captioners create captions for a wide range of live broadcasts on network, cable, syndication, and pay-per-view services. All programs on prime-time schedules of the three major commercial networks are now captioned, many by real-time captioners.

Real-time captions are generated within seconds after a word is spoken. They are made possible by highly skilled court reporters who receive months of specialized retraining to become first-class real-time captioners.

The Job

The refined skills of real-time captioners are called upon every day to bring the latest news, sports, and entertainment to a diverse group consisting not only of the deaf and hard-of-hearing, but young children learning to read and those learning English as a second language. While captioning a live program, meeting, or other event may seem rather straightforward on the surface, there is a great deal of work, anxiety, and preparation that goes into ensuring that the words appearing on screen come out as smoothly and effortlessly as possible. Real-time captioning requires much dexterity and discipline to be able to reach the higher speeds required—250 words a minute—and good brain-to-hand coordination to get it all down quickly and accurately.

There is also much preparation work that must be done by real-time captioners before they can caption a live television broadcast. It takes about one-and-a-half to two hours to prepare for an average news broadcast, using preparation materials obtained from the broadcaster and the captioner's own research. (Special broadcasts like holiday parades, the Super Bowl, or the Olympics can take days or even weeks of preparation.) Captioners call this pre-show preparation "dictionary-building."

Captioners working for established captioning houses will usually have access to all types of reference materials—everything from *Star Stats Who's Who in Hollywood* to the *Congressional Staff Directory*. Captioners working on their own will want to think about what kinds of materials to include in their own libraries.

Real-time captioners prepare for a job by going through resource materials to find words that might come up during a broadcast, then develop "briefs" or steno codes that they will use to "write" these words when the words come up during the broadcast. It is important that captioners test all the briefs developed for complicated names to make sure they are translating properly. Because captioners will hear names and words during the broadcast that they have not prepared dictionary entries for, they must learn to "write around" the actual words and listen for titles. In this way captioners can write "The former Secretary of State" instead of "Henry Kissinger," for example.

While striving to keep them to a minimum, captioners will occasionally make mistakes that go out over the air. For example, in real-time captioning, the phrase "Olympic tryouts," which would require the captioner to type five key strokes on a stenotype machine, might come out (and actually did) as "old limp pig tryouts" if strokes are entered that the computer cannot match correctly.

CART reporters also work in classroom settings, where they might be seen with a notebook computer and steno keyboard, sitting next to a deaf person. CART reporters write down everything that happens, making sure the notebook computer screen is turned so the deaf person can see it. To help the client better understand what is going on, they may paraphrase or interpret the proceedings, not just create a verbatim record, as in a courtroom. Real-time reporters can also cover meetings, with captions shown on large projection screens. Additionally, computer technology allows highly skilled court reporters to provide real-time captioning in the courtroom, which has great value for large numbers of deaf or hard-of-hearing judges, attorneys, and litigants, or those who have difficulty understanding English. Also, judges and attorneys can scroll back to earlier statements during the trial and mark text for later reference.

One major difference between real-time captioning for television broadcast and other live-display settings and verbatim reporting, as is frequently done in courtrooms and lawyers' offices, is that captioning's main purpose is to let the viewer who is deaf or hard-of-hearing understand the story being told on the screen. It is not enough to listen only for the phonetic strokes; the real-time captioner must also listen for context.

Sheri Smargon works for a captioning company in Tampa, Florida. "I caption the news for about 12 different stations around the country," she says. "My company has more, but I have regular cities that I'm usually responsible for. The news consists of anything from a half-hour program to two hours of straight news. I also caption NBA and MLB games." While captioning, Smargon doesn't receive a TV picture. "I get an audio feed only, so I just write what I hear," she says. "Hockey games were the hardest...everyone's name sounds alike!"

Before beginning even limited on-air captioning, captioner trainees must spend at least three to six months in training, eight hours a day, five days a week, and up to one year of real-time captioning before doing certain specialized programming. As a vital part of the production team, captioners must also become intimately familiar with the programs they are captioning to know what to expect and to anticipate the unexpected.

A typical day for a captioner trainee would include preparing for a practice broadcast by creating a job dictionary, then writing that practice broadcast for supervisors, who would make suggestions as to conflict resolution, editing, brief form, style, and format. Later, the trainee would review the broadcast and make the necessary dictionary entries. Trainees would sit in on a variety of broadcasts with more experienced captioners.

Real-time captioning for television is generally performed in a production control room, equipped with several television sets and networked computer systems, giving the environment a high-tech look and feel. Sometimes, one captioner will write a show alone; sometimes two captioners will share a show, depending on whether there are commercials or not. No captioner can maintain a high accuracy level without taking regular breaks. On a show with no commercials, two captioners would typically switch back and forth about every 10 minutes.

As a show gets closer to air, the environment in the control room becomes tense, as the real-time captioner scrambles to get last-minute information in the computer. Then a deep breath, and the countdown begins . . . "Good evening, I'm Peter Jennings."

The captioner strokes the steno keys while listening to the live broadcast, transcribing the broadcast accurately while inserting correct punctuation and other symbols. (Double arrows at the beginning of a sentence indicate that a new speaker is speaking.) Those strokes are converted to electronic impulses, which travel through a cable to the computer. The steno strokes are matched with the correct entries on the captioner's personal dictionary. That data is then sent by modem to the broadcast site, where it gets added to the broadcaster's video signal. Within two to three seconds, people across the country can see those captions—if they have televisions with a built-in decoder chip or a set with a decoder connected to it.

Some kinds of captioning can be done from home, mainly broadcasts for local television stations. The equipment needed (which may be provided by the employer) includes a computer, modem, steno machine, and the appropriate software. Captioners may even choose to work for companies that specialize in producing captions remotely, with just an audio feed, thereby allowing more home-based operations. Getting started in the business, however, usually requires an on-site presence, until confidence and trust is established. Obviously, live events that are not broadcast will require a real-time captioner on site.

Requirements

HIGH SCHOOL
You should take typing and computer courses to increase keyboard speed and accuracy and to develop an understanding of word processing programs. Because you'll be working with a variety of news, sports, and entertainment programs, you should keep up on current events by taking journalism, social studies, and government courses. English composition and speech classes can help you develop your vocabulary and grammar skills.

POSTSECONDARY TRAINING
You should first complete training to become a court and conference reporter (stenographer), which takes anywhere from two to four years. An associate's or bachelor's degree in court and conference reporting, or satisfactory completion of other two-year

equivalent programs, is usually required. Because of the additional training needed to learn computer and English grammar skills, some of the formerly two-year programs have gone to three. In fact, many real-time reporters and their employers believe that additional formal education in the arts and sciences is needed to perform the work properly and to adapt to the swift technological changes taking place. They are urging the National Court Reporters Association (NCRA), to which most captioners and other reporters belong, to require a bachelor's degree for entry into the court reporting profession, which would extend to captioning as well. A few four-year college programs already exist to allow students a well-rounded background. A degree in English (or the primary language in which captioning will be done) or linguistics would be helpful. Others argue, however, that while a formal education is beneficial, many court reporters who never earned a four-year degree are working successfully with high skill levels.

Even after graduating from court reporting school, you will have to undergo more specialized training, during which you'll hone your reporting skills to achieve the proficiency needed to create broadcast-quality captions.

CERTIFICATION OR LICENSING

Typically, the reporter considering real-time captioning work has passed the Registered Professional Reporter (RPR) exam given by the NCRA, or a comparable state certification exam. Potential employers may even require it. The skills and knowledge needed to pass this exam are similar to those required for captioning, though not as stringent. Anyone capable of doing broadcast-quality captioning work can easily get RPR certification.

OTHER REQUIREMENTS

You should have extreme proficiency in machine shorthand skills and an ability to perform under pressure. Familiarity with CAT systems is usually preferred, as is previous court or field reporting experience. It generally takes several years of court reporting experience to be able to take and transcribe complex testimony with the high levels of speed and accuracy that real-time captioning demands.

Real-time captioners must also possess an incredible amount of concentration. Besides typing accurately at speeds of 200 to 250 words a minute to keep up with the fastest natural speakers, they must also anticipate commercial breaks so as not to cut off captions in mid-sentence, insert appropriate punctuation marks and symbols, and watch their own translation closely to correct any problems on the spot.

"I try to stay informed about what's going on in the world," Sheri Smargon says, "not just in the news. It helps to know that 'Eminem' has a new record, as well as to know that Kosovo is a province, not a city."

Exploring

Although some core classes on captioning technology are being injected into court reporting and other stenographic curricula around the country, it is still a "hit or miss" situation, with many schools simply intimidated by the new technology. Good programs exist, however, that are providing beneficial exposure and actually working with local TV stations and area colleges to provide both news captioning and real-timing or steno interpreting in the classroom for deaf students and those with disabilities.

A smart way to prepare for real-time captioning, according to a real-time captioner who hires new graduates for a captioning company, is to practice by transcribing or writing newspaper articles or those from news magazines. Along with helping to build vocabulary skills, this exercise enables you to focus on conflict resolution by seeing the word in print, helps to familiarize you with difficult foreign names and words, and increases awareness of current events, both national and international.

While honing your skills, you may also get good exposure by working with local organizations, such as the Association of Late Deafened Adults, Self-Help for Hard of Hearing Persons, the National Association of the Deaf, the Alexander Graham Bell Association for the Deaf and Hard of Hearing, and other nonprofit groups that might eventually need captioning services. Although the pay will not be as high as it would at a captioning house, the job satisfaction level will be high. It is good to keep in mind that while the major captioning companies do sometimes hire people with little or no training for internships or on-the-job training, there is no substitute for experience.

Employers

Captioners are employed primarily by captioning companies such as NCI and VITAC. These companies contract with broadcasters and production companies to caption live and recorded events. Captioners either work as full-time employees for captioning companies or as freelancers (that is, independent contractors).

Starting Out

You should seek employment at one of the few large captioning companies in the country or contact station managers at your local television stations to inquire about real-time captioning positions. As with many other businesses, the best approach may be simply to start calling the leading companies in the field and the local companies and see who is hiring. Gallaudet University (http://www2.gallaudet.edu) in Washington, DC, puts out a list of captioning companies.

Before securing a real-time captioning position, you may have to "audition" as part of a pre-interview screening process that involves preparing raw steno notes from a sample tape-recorded program. The notes are then analyzed, with employment consideration based on the results of the evaluation and job experience. A good way to prepare for employment evaluation is to practice on the kind of material you wish to caption and to offer to demonstrate your skills.

Advancement

Advancement for a real-time captioner is dependent upon performance, with salary increases and promotions to more responsible positions awarded with greater proficiency and tenure. Skilled real-time captioners may advance to supervisory positions.

Earnings

Earning power for real-time captioners is dependent upon many variables and is often region-specific and a product of "what the market will bear." In large captioning organizations, real-time captioners can make anywhere from $28,000 for a recent graduate in training to $65,000 or even higher for those experienced and tireless workers who always volunteer for extra hours, overflow work, etc., and who are capable of captioning all kinds of programming. Trainee salaries increase once the captioner goes on the air. According to *The O*Net Dictionary of Occupational Titles*, the mean annual earnings of all captioners was approximately $32,920.

Salaries for real-time captioners are often in line with salaries for court reporters. According to an NCRA survey of its members, average annual earnings for court reporters were $61,830 in 1999.

A fringe benefit of working for a captioning agency for most reporters (particularly students just out of school) is that such agencies generally provide all the equipment, which would cost approximately $15,000. Large captioning organizations also offer benefits, such as vacation and health insurance, likely to be provided at a courthouse for court reporters but not at a freelance firm of deposition reporters, for instance.

Work Environment

Real-time captioning for television broadcast is not a nine-to-five job. While many reporting jobs require erratic hours, broadcast captioning is done seven days a week, around the clock. Real-time captioners producing captions for television broadcast will likely work nights, weekends, or holidays, as directed. Shows can air at 5:30 in the morning, at midnight on a Saturday night, or during Thanksgiving dinner.

Given the irregularity of TV schedules, several shifts are needed to cover programming hours scheduled throughout the day. It is imperative that captioners be flexible and dependable and that they not get fatigued, so they can maintain high accuracy levels. How many hours a day a captioner is on the air depends on the level of experience. If new to the air, captioners may do only one or two shows a day, as it takes longer to prepare for a broadcast and review the result in the beginning. An experienced captioner may be on the air three to five hours a day, writing short programs or a news broadcast or sporting event. At least, in the broadcast setting, real-time captioners do not have to produce transcripts, which eliminates the long hours that go along with that aspect of reporting.

Real-time captioning work can be physically demanding. Along with suffering the mental stress of performing in a live environment, real-time captioners may also be subject to repetitive stress injury, a prevalent industrial hazard for those who perform repeated motions in their daily work. Carpal tunnel, a type of repetitive stress injury, sometimes afflicts real-time captioners after several years. It can cause prickling sensation or numbness in the hand and sometimes a partial loss of function.

"Captioning and real-timing are totally different from regular court reporting," Sheri Smargon says. "You have to be a certain kind of person to real-time. I find captioning challenging and rewarding and fun, usually."

Outlook

The NCRA reports a decline in enrollment in court reporting schools. This may be because of the development of voice and speech systems—the computer programs that automatically convert speech to written text. However, there are no current systems that can accurately handle multiple speakers, and it's unlikely that such technology will exist in the near future. Therefore, captioners and court reporters will be in high demand for years to come. New requirements by the Federal Communications Commission are also increasing demands for captioners. The Telecommunications Act of 1996, for example, requires that 95 percent of all new programming on television must be captioned by 2006.

Digital TV (DTV) will also make captioning more desirable and useful to more people, thereby increasing demand for captioners. DTV enhancements will allow viewers with poor vision to adjust text-size, styles, and fonts. DTV will also allow for more non-English letters, as well as more information transmitted per minute.

Captioners should focus first on the area where they want to live and work. To caption area news or city council meetings in a local area or do conventions in a large hotel, captioners must first obtain some costly supplies. These include a laptop or notebook computer, a compatible steno writer, cables, modem, and captioning software. Captioners may also need a character generator to project onto a large convention screen.

Captioners should learn the basic real-time skills that will enable them to do any live translating or interpreting on the computer. With such skills, they will be eligible for a variety of positions, including working in a computer-integrated courtroom; taking real-time depositions for attorneys; providing accompanying litigation support, such as key word indexing; real-timing or captioning in the classroom; or doing broadcast captioning. The future looks great for those who qualify themselves to perform real-time translation.

Other opportunities for the real-time captioner include working with hospitals that specialize in cochlear implants. For late-deafened adults who learned English before sign language, if they learned to sign at all, captions provide a far greater comprehension level. Additionally, some local news stations across the country are working to expand and improve the quality of their local captioning capabilities, providing yet another source of potential employment for the real-time captioner.

For More Information

The NCI Web site features historical information, a list of captioning terms, and employment information.
NATIONAL CAPTIONING INSTITUTE (NCI)
1900 Gallows Road, Suite 3000
Vienna, VA 22182
Tel: 703-917-9878
Web: http://www.ncicap.org

Visit the NCRA Web site for extensive career and certification information, as well as information about technology, education programs, and access to the NCRA monthly magazine.
NATIONAL COURT REPORTERS ASSOCIATION (NCRA)
8224 Old Courthouse Road
Vienna, VA 22182-3808
Tel: 800-272-6272
Web: http://www.ncraonline.org

Reporters

Overview

Reporters are the foot soldiers for newspapers, magazines, and television and radio broadcast companies. They gather and analyze information about current events and write stories for publication or for broadcasting. News analysts, reporters, and correspondents hold about 78,000 jobs in the United States.

History

Broadcast technology developed in the 20th century enabled people to reach large audiences around the world instantly, changing forever the way we communicate.

Small radio shows started in 1910. Ten years later, two commercial radio stations went on the air, and by 1921, a dozen local stations were broadcasting. The first network radio broadcast (more than one station sharing a broadcast) was of the 1922 World Series. By 1926, stations across the country were linked together to form the National Broadcasting Company (NBC). Four years later, the first radio broadcast was transmitted around the world.

Although the advent of television changed the kind of programming available on the radio (from comedy, drama, and news programs to radio's current schedule of music, phone-in talk shows, and news updates), there has been a steady growth in the number of radio stations in the United States. According to the Federal Communications Commission, there were 13,012 radio stations in the United States in 2001.

Modern television developed from experiments with electricity and vacuum tubes in the mid-1800s, but it was not until 1939, when President Franklin Roosevelt used television to open the New York World's Fair, that the public realized the power of television as a means of communication. Several stations went on the air shortly after this demonstration and successfully televised professional baseball games, college football games, and the Republican and Democratic conventions of 1940. The onset of World War II limited the further development of television until after the war was over.

Since television's strength is the immediacy with which it can present information, news programs became the foundation of regular programming. *Meet the Press* premiered in 1947, followed by nightly newscasts in 1948. The industry expanded rapidly in the 1950s. The Federal Communications Commission lifted a freeze on the processing of station applications, and the number of commercial stations grew steadily, from 120 in 1953 to over 1,600 stations today.

In today's complex world, with the public hungry for news as it occurs, reporters and correspondents are involved in all media—from newspapers and other print publications to radio and television. Today, with the advent of the Internet, many newspapers and television and radio stations are going online, causing many reporters to become active participants on the information superhighway.

The Job

Reporters collect information on newsworthy events and prepare stories for newspaper or magazine publication or for radio or television broadcast. The stories may simply provide information about local, state, or national events, or they may present opposing points of view on issues of current interest. In this latter capacity, the press plays an important role in monitoring the actions of public officials and others in positions of power.

Stories may originate as an assignment from an editor or as the result of a lead or news tip. Good reporters are always on the lookout for good story ideas. To cover a story, they gather and verify facts by interviewing people involved in or related to the event, examining documents and public records, observing events as they happen, and researching relevant background information.

Reporters generally take notes or use a tape recorder as they collect information and write their stories once they return to their offices. In order to meet a deadline, they may have to telephone the stories to *rewriters*, who write or transcribe the stories for them. After the facts have been gathered and verified, the reporters transcribe their notes, organize their material, and determine what emphasis, or angle, to give the news. The story is then written to meet prescribed standards of editorial style and format.

The basic functions of reporters are to observe events objectively and impartially, record them accurately, and explain what the news means in a larger, societal context. Within this framework, there are several types of reporters.

The most basic is the *news reporter.* This job sometimes involves covering a beat, such as the police station, courthouse, or school system. It may involve receiving general assignments, such as a story about an unusual occurrence or an obituary of a community leader. Large daily papers or large radio and television stations may assign teams of reporters to investigate social, economic, or political events and conditions.

Many reporters specialize in one type of story, either because they have a particular interest in the subject or because they have acquired the expertise to analyze and interpret news in that particular area. *Topical reporters* cover stories for a specific department, such as medicine, politics, foreign affairs, sports, consumer affairs, finance, science, business, education, labor, or religion. They sometimes write features explaining the history that has led up to certain events in the field they cover. *Feature writers and reporters* generally write longer, broader stories than news reporters, usually on more upbeat subjects, such as fashion, art, theater, travel, and social events. They may write about trends, for example, or profile local celebrities. *Editorial writers, commentators,* and *syndicated news columnists* present viewpoints that, although based on a thorough knowledge, are opinions on topics of popular interest. *Columnists* write under a byline and usually specialize in a particular subject, such as politics or government activities. *Critics* review restaurants, books, works of art, movies, plays, musical performances, and other cultural events.

Specializing allows reporters to focus their efforts, talent, and knowledge on one area of expertise. It also allows them more opportunities to develop deeper relationships with contacts and sources necessary to gain access to the news.

Correspondents report events in locations distant from their home offices. They may report news by mail, telephone, fax, or computer from rural areas, large cities throughout the United States, or countries. Many large newspapers, magazines, and broadcast companies have one correspondent who is responsible for covering all the news for the foreign city or country where they are based.

Requirements

HIGH SCHOOL
High school courses that provide a firm foundation for a career as reporter include English, journalism, social studies, speech, typing, and computer science.

POSTSECONDARY TRAINING
A bachelor's degree is essential for aspiring reporters. Graduate degrees give students a great advantage over those entering the field with lesser degrees. Most editors prefer applicants with degrees in journalism because their studies include liberal arts courses as well as professional training in journalism. Some editors consider it sufficient for a reporter to have a good general education from a liberal arts college. Others prefer applicants with an undergraduate degree in liberal arts and a master's degree in journalism.

More than 400 colleges offer programs in journalism leading to a bachelor's degree. In these schools, around three-fourths of a student's time is devoted to a liberal education and one-fourth to the professional study of journalism, with required courses such as introductory mass media, basic reporting and copy editing, history of journalism, and press law and ethics. Students are encouraged to select other journalism courses according to their specific interests.

Journalism courses and programs are also offered by many community and junior colleges. Graduates of these programs are prepared to go to work directly as general assignment reporters, but they may encounter difficulty when competing with graduates of four-year programs. Credit earned in community and junior colleges may be transferable to four-year programs in journalism at other colleges and universities. Journalism training may also be obtained in the armed forces. Names and addresses of newspapers and a list of journalism schools and departments are published in the annual *Editor & Publisher International Year Book: The Encyclopedia of the Newspaper Industry,* which is available for reference in most public libraries and newspaper offices.

A master's degree in journalism may be earned at approximately 120 schools, and a doctorate at about 35 schools. Graduate degrees may prepare students specifically for careers in news or as journalism teachers, researchers, and theorists or for jobs in advertising or public relations.

A reporter's liberal arts training should include courses in English (with an emphasis on writing), sociology, political science, economics, history, psychology, business, speech, and computer science. Knowledge of foreign languages is also useful. To be a reporter in a specialized field, such as science or finance, requires concentrated course work in that area.

OTHER REQUIREMENTS

A crucial requirement for reporters is typing skill. Reporters type their stories using word processing programs. Although not essential, a knowledge of shorthand or speedwriting makes note taking easier, and an acquaintance with news photography is an asset.

Reporters must be inquisitive, aggressive, persistent, and detail-oriented. They must enjoy interaction with people of various races, cultures, religions, economic levels, and social statuses.

Exploring

You can explore a career as a reporter in a number of ways. You can talk to reporters and editors at local newspapers and radio and TV stations. You can interview the admissions counselor at the school of journalism closest to your home.

In addition to taking courses in English, journalism, social studies, speech, computer science, and typing, high school students can acquire practical experience by working on school newspapers or on a church, synagogue, or mosque newsletter. Part-time and summer jobs on newspapers provide invaluable experience to the aspiring reporter.

College students can develop their reporting skills in the laboratory courses or workshops that are part of the journalism curriculum. College students might also accept jobs as campus correspondents for selected newspapers. People who work as part-time reporters covering news in a particular area of a community are known as stringers and are paid only for those stories that are printed.

More than 3,000 journalism scholarships, fellowships, and assistantships are offered by universities, newspapers, foundations, and professional organizations to college students. Many newspapers and magazines offer summer internships to journalism students to provide them with practical experience in a variety of basic reporting and editing duties. Students who successfully complete internships are usually placed in jobs more quickly upon graduation than those without such experience.

Employers

Of the approximately 78,000 reporters and correspondents employed in the United States, nearly 50 percent work for newspapers of all sizes. About 28 percent work in radio and television broadcasting. The rest are employed by wire services and magazines.

Starting Out

Jobs in this field may be obtained through college placement offices or by applying directly to the personnel departments of individual employers. Applicants with some practical experience will have an advantage; they should be prepared to present a portfolio of material they wrote as volunteer or part-time reporters or other writing samples.

Most journalism school graduates start out as general assignment reporters or copy editors for small publications. A few outstanding journalism graduates may be hired by large city newspapers or national magazines. They are trained on the job. But they are the exception, as large employers usually require several years' experience. As a rule, novice reporters cover routine assignments, such as reporting on civic and club meetings, writing obituaries, or summarizing speeches. As reporters become more skilled, they are assigned to more important events or to a regular beat, or they may specialize in a particular field.

Advancement

Reporters may advance by moving to larger newspapers, press services, or radio and television markets, but competition for such positions is unusually keen. Many highly qualified reporters apply for these jobs every year.

A select number of reporters eventually become columnists, correspondents, editorial writers, editors, or top executives. These important and influential positions represent the top of the field, and competition is strong for them.

Many reporters transfer the contacts and knowledge developed in newspaper reporting to related fields, such as public relations, advertising, or preparing copy for radio and television news programs.

Earnings

There are great variations in the earnings of reporters. Salaries are related to experience, kind of employer for which the reporter works, geographical location, and whether the reporter is covered by a contract negotiated by the Newspaper Guild.

According to the Newspaper Guild, the average top minimum salary for reporters with about five years' experience was $41,400 in 2000. Salaries ranged from $20,150 to $65,528.

The U.S. Department of Labor reports median annual earnings of $29,110 for news analysts, reporters, and correspondents in 2000. Salaries ranged from less than $16,540 to more than $69,300. Reporters and correspondents who worked in radio and television broadcasting had median annual earnings of $33,550, compared to median earnings of $26,900 for those who worked in newspapers.

According to a 1999 survey conducted by the National Association of Broadcasters and the Broadcast Cable Financial Management Association, the annual average salary for television news reporters was $33,700. According to the 2001 survey, the annual average salary, including bonuses, was $55,100 for radio news reporters and $53,300 for sportscasters in radio broadcasting.

Work Environment

Reporters work under a great deal of pressure in settings that differ from the typical business office. Their jobs generally require a five-day, 35- to 40-hour week, but overtime and irregular schedules are very common. Reporters employed by morning papers start work in the late afternoon and finish around midnight, while those on afternoon or evening papers start early in the morning and work until early or midafternoon. Foreign correspondents often work late at night to send the news to their papers in time to meet printing deadlines.

Travel is often required in this occupation, and some assignments may be dangerous, such as covering wars, political uprisings, fires, floods, and other events of a volatile nature.

Outlook

Employment for reporters and correspondents through 2010 is expected to grow more slowly than the average for all occupations, according to the *Occupational Outlook Handbook*. The U.S. Bureau of Labor Statistics projects that the number of employed reporters and correspondents will decline by about 4.4 percent between 2000 and 2010. While the number of self-employed reporters and correspondents is expected to grow, newspaper jobs are expected to decrease because of mergers, consolidations, and closures in the radio, television, and newspaper industries.

A significant number of jobs will be provided by magazines and in radio and television broadcasting, but the major news magazines and larger broadcasting stations generally prefer experienced reporters. For beginning correspondents, small stations with local news broadcasts will continue to replace staff who move on to larger stations or leave the business. Network hiring has been cut drastically in the past few years and will probably continue to decline.

Some employment growth is expected for reporters in online newspapers and magazines.

Poor economic conditions do not drastically affect the employment of reporters and correspondents. Their numbers are not severely cut back even during a downturn; instead, employers forced to reduce expenditures will suspend new hiring.

For More Information

This organization provides general educational information on all areas of journalism (newspapers, magazines, television, and radio).
ASSOCIATION FOR EDUCATION IN JOURNALISM AND MASS COMMUNICATION
234 Outlet Pointe Boulevard
Columbia, SC 29210-5667
Tel: 803-798-0271
Email: aejmchq@aejmc.org
Web: http://www.aejmc.org

Contact BEA for scholarship information and a list of schools offering degrees in broadcasting. Visit their Web site to sign up for a free monthly email newsletter that contains useful information about broadcast education and the broadcasting industry:
BROADCAST EDUCATION ASSOCIATION (BEA)
1771 N Street, NW
Washington, DC 20036-2891
Tel: 202-429-5354
Web: http://www.beaweb.org

To receive a copy of The Journalist's Road to Success, *which lists schools offering degrees in news-editorial and financial aid to those interested in print journalism, contact or visit the following Web site:*
DOW JONES NEWSPAPER FUND
4300 Route One North
South Brunswick, NJ 08852
Tel: 609-452-2820
Email: newsfund@wsj.dowjones.com
Web: http://www.dowjones.com/newsfund

For job listings and union information, contact:
NATIONAL ASSOCIATION OF BROADCAST EMPLOYEES AND TECHNICIANS
501 Third Street, NW, 8th Floor
Washington, DC 20001
Tel: 202-434-1254
Email: nabet@nabetcwa.org
Web: http://union.nabetcwa.org

For information on scholarships and student membership, contact:
NATIONAL ASSOCIATION OF FARM BROADCASTERS
26 East Exchange Street, Suite 307
St. Paul, MN 55101
Tel: 651-224-0508
Email: info@nafb.com
Web: http://www.nafb.com

For statistics and contact information for the cable industry, visit the NCTA Web site or contact:
NATIONAL CABLE AND TELECOMMUNICATIONS ASSOCIATION (NCTA)
1724 Massachusetts Avenue, NW
Washington, DC 20036
Tel: 202-775-3550
Web: http://www.ncta.com

For information on careers in newspapers and industry facts and figures, contact:
NEWSPAPER ASSOCIATION OF AMERICA
1921 Gallows Road, Suite 600
Vienna, VA 22182-3900
Tel: 703-902-1600
Web: http://www.naa.org

Sports Broadcasters and Announcers

Overview

Sports broadcasters for radio and television stations select, write, and deliver footage of current sports news for the sports segment of radio and television news broadcasts or for specific sports events, channels, or shows. They may provide pre- and postgame coverage of sports events, including interviews with coaches and athletes, as well as play-by-play coverage during the game or event.

Sports announcers are the official voices of the teams. At home games it is the sports announcer who makes pregame announcements, introduces the players in the starting lineups, and keeps the spectators in the stadium or arena abreast of the details of the game by announcing such things as fouls, substitutions, and goals, and who is making them.

History

Radio signals, first transmitted by Guglielmo Marconi in 1895, led to early experimentation with broadcasting in the years preceding World War I. After the war began, however, a ban on amateur, non-military radio broadcasts delayed radio's acceptance. In 1919, when the ban was lifted, hundreds of amateur stations sprang up. By 1922, 500 were licensed by the government. Codes and domestic broadcast wavelengths were assigned by the government, which created a traffic jam of aerial signals. Eventually, more powerful stations were permitted to broadcast at a higher wavelength, provided these stations only broadcast live music. This move by the government quickly brought entertainment from large, urban areas to the small towns and rural areas that characterized most of the United States at the time.

In the early days of radio broadcasts, anyone who operated the station would read, usually verbatim, news stories from the day's paper. Quickly, station managers realized that the station's "voice" needed as much charisma and flair as possible. Announcers and journalists with good speaking voices were hired. With the arrival of television, many of those who worked in radio broadcasting moved to this new medium.

Corporate-sponsored radio stations weren't long in coming; Westinghouse Corporation and American Telephone and Telegraph (AT&T) raced to enter the market. Westinghouse engineer Frank Conrad received a license for what is viewed as the first modern radio station, KDKA, in Pittsburgh, Pennsylvania. KDKA broadcast music programs, the 1920 presidential election, and sports events. The next year, Westinghouse began to sell radio sets for as little as $25. By 1924, the radio-listening public numbered 20 million.

Meanwhile, as early as 1929 a Soviet immigrant employed by Westinghouse, Vladimir Kosma Zworykin, was experimenting with visual images to create an all-electronic television system. By 1939 the system was demonstrated at the New York's World Fair with none other than President Franklin D. Roosevelt speaking before the camera. World War II and battles over government regulation and AM and FM frequencies interrupted the introduction of television to the American public, but by 1944, the government had determined specific frequencies for both FM radio and television.

In 1946, the number of television sets in use was 6,000; by 1951, the number had risen to an astonishing 12 million sets. Unknowingly, the stage had been set for a battle between radio and television. In the ensuing years, expert after expert predicted the demise of radio. The popularity of television, its soap operas, family dramas, and game shows, was believed by nearly everyone to be too strong a competitor for the old-fashioned, sound-only aspect of radio. The experts proved wrong; radio flourished well into the 1990s, when the industry experienced some cutbacks in the number of stations and broadcast hours because of recession.

The national radio networks of the early days are gone, but satellites allow local stations to broadcast network shows anywhere with the equipment to receive the satellite link. The development of filmed and videotaped television, cable and satellite transmissions, broadcasting deregulation, and an international market through direct broadcast satellite systems has drastically changed the face and future of both radio and television.

Today's sports broadcasters in radio and television have all these technological tools and more at their fingertips. Want to see an instant replay of the game-winning three-point shot by Kobe Bryant? As the sportscaster describes it, a technician is playing it back for the viewing public. Have to travel to Costa Rica for a business trip, but hate to miss that Yankees' game? No problem. A sportscaster is giving the play-by-play to an AM network station that is, in turn, sending it via satellite to a Costa Rican client-station—or, the game may even be broadcast via the Internet!

The Job

One of the primary jobs of most sportscasters for both radio and television stations is to determine what sports news to carry during a news segment. The sportscaster begins working on the first broadcast by reading the sports-related clippings that come in over the various news wire services, such as the Associated Press and United Press International. To follow up on one of these stories, the sportscaster might telephone several contacts—a coach, a scout, an athlete—to see if he can get a comment or more information. The sportscaster also might want to prepare a list of upcoming games, matches, and other sports events. Athletes often make public appearances for charity events and the sportscaster might want to include a mention of the charity and the participating athlete or athletes.

After deciding which stories to cover and the lineup of the stories that will be featured in the first of the day's broadcasts, sportscasters then review any audio or video clips that will accompany the various stories. Sportscasters working for radio stations choose audio clips, usually interviews, that augment the piece of news they will read. Sportscasters working for television stations look for video footage—the best 10 seconds of this game or that play—to demonstrate why a certain team lost or won. Sometimes sportscasters choose footage that is humorous or poignant to illustrate the point of the news item.

After they decide which audio or video segments to use, sportscasters then work with sound or video editors to edit the data into a reel or video, or they edit the footage into a tape themselves. In either case, the finished product will be handed over to the news director or producer with a script detailing when it should play. The news producer or director will make certain that the reel or video comes in on cue during the broadcast.

Frequently, a sportscaster will make brief appearances at local sports events to interview coaches and players before and after the game, and sometimes during breaks in the action. These interviews, as well as any footage of the game that the station's camera crews obtain, are then added to the stock from which sportscasters choose for their segments.

Usually, the main broadcast for both radio and television sportscasters is the late evening broadcast following the evening's scheduled programming. This is when most of the major league sports events have concluded, the statistics for the game are released, and final, official scores are reported. Any changes that have occurred since the day's first sports broadcast are updated and new footage or sound bites are added. The final newscast for a television sportscaster will most likely include highlights from the day's sports events, especially dramatic shots of the most impressive or winning points scored.

Increasingly, in televised sports news the emphasis is on image. Often sportscasters—like other newscasters—are only on camera for several seconds at a time, but their voices continue over the videotape that highlights unique moments in different games.

For many sportscasters who work in television, preparing the daily sportscasts is their main job and takes up most of their time. For others, especially sportscasters who work in radio, delivering a play-by-play broadcast of particular sports events is the main focus of their job. These men and women use their knowledge of the game or sport to create with words a visual picture of the game, as it is happening, for radio listeners. The most common sports for which sportscasters deliver play-by-play broadcasts are baseball, basketball, football, and hockey. A few sportscasters broadcast horse races from the race track and sometimes these broadcasts are carried by off-track betting facilities.

Sportscasters who give the play-by-play for a basketball game, for example, usually arrive an hour or so before the start of the game. Often, they have a pregame show that features interviews with, and a statistical review of, the competing teams and athletes. To broadcast a basketball game, sportscasters sit courtside in a special media section so that they can see the action up close. During football, baseball, and hockey games sportscasters usually sit in one of the nearby media boxes. Throughout the game sportscasters narrate each play for radio listeners using rapid, precise, and lively descriptions. During time-outs, half-times, or other breaks in play, sportscasters might deliver their own running commentaries of the game, the players' performances, and the coaching.

Although some skills are advantageous to both aspects of the job, the sportscaster who specializes in play-by-play broadcasts needs to have an excellent mastery of the rules, players, and statistics of a sport, as well as the hand signals used by officials to regulate the flow of a game. Some sportscasters provide play-by-play broadcasts for several different teams or sports, from college to professional levels, requiring them to know more than one sport or team well.

Some sportscasters—often former athletes or established sports personalities—combine the two aspects of the job. They act as anchors or co-anchors for sports shows and give some play-by-play commentary while also providing their television or radio audience with statistics and general updates.

In a related job, sports announcers provide spectators with public address announcements before and during a sports event. For this job, sportscasters must remain utterly neutral, simply delivering the facts—goals scored, numbers of fouls, or a time-out taken. Sports announcers may be sportscasters or they may be professional announcers or emcees who make their living recording voice-overs for radio and television commercials and for large corporations or department stores.

Sports announcers usually give the lineups for games, provide player names and numbers during specific times in a contest, make public announcements during time-outs and pauses in play, and generally keep the crowd involved in the event (especially in baseball). Baseball announcers may try to rally the crowd or start the crowd singing or doing the wave.

Requirements

HIGH SCHOOL

Graduating from high school is an important first step on the road to becoming a sports broadcaster or announcer. While in school, take classes that will allow you to work on your speaking and writing skills. Classes in speech, English, and foreign languages, such as Spanish and French, will be helpful.

POSTSECONDARY TRAINING

Educational requirements for sportscasting positions vary, depending on the position. Competition for radio and television sports broadcasting positions is especially fierce, so any added edge can make the difference.

Television sportscasters who deliver the news in sports usually have bachelor's degrees in communications or journalism, although personality, charisma, and overall on-camera appearance is so important to ratings that station executives often pay closer attention to the taped auditions they receive from prospective sportscasters than to the items on resumes. If you are interested in pursuing a career in sports broadcasting, keep in mind that the industry is finicky and subjective about looks and charisma, so you should continue to prepare yourself for the job by learning a sport inside and out, developing valuable contacts in the field through internships and part-time or volunteer jobs, and earning a degree in journalism or communications.

It isn't as crucial for sportscasters who deliver play-by-play broadcasts for radio stations to have the journalistic skills that a television sportscaster has, although good interviewing skills are essential. Instead, they need excellent verbal skills, a daunting command of the sport or sports that they will be broadcasting, and a familiarity with the competing players, coaches, and team histories. To draw a complete picture for their listeners, sportscasters often reach back into history for an interesting detail or statistic, so a good memory for statistics and trivia and a knowledge of sports history is helpful, too.

OTHER REQUIREMENTS

A nice speaking voice, excellent verbal and interviewing skills, a pleasant appearance, a solid command of sports in general as well as in-depth knowledge of the most popular sports (football, hockey, basketball, and baseball), and an outgoing personality are all necessary for a successful career in sportscasting.

Sports announcers need to have strong voices, excellent grammar and English usage, a pleasant appearance, and the ability to ad-lib if and when it is necessary. A solid knowledge of the sport is essential.

Exploring

High school and college students have many opportunities to investigate this career choice, but the most obvious way is to participate in a sport. By learning a sport inside and out, you can gain valuable insight into the movements and techniques that, as a sportscaster, you will be describing. In addition, firsthand experience and a love of the sport itself makes it easier to remember interesting trivia related to the sport as well as the names and numbers of the pros who play it.

If you do not have the coordination or skill for the sport itself, you can volunteer to help out with the team by shagging balls, running drills, or keeping statistics. The latter is perhaps the best way to learn the percentages and personal athletic histories of athletes.

An excellent way to develop the necessary communications skills is to take a journalism course, join the school's speech or debate team, deliver the morning announcements, deejay on the school radio station, or volunteer at a local radio station or cable television station.

John Earnhardt from the National Association of Broadcasters has this advice: "Write about your school's sports teams for your school newspaper or hometown newspaper and read, read, read about sports. Knowledge about the area you are interested in reporting about is the best tool for success. It is also necessary to be able to express yourself well through the spoken word. Speaking before an audience can be the best practice for speaking before the camera or on a microphone."

Finally, many aspiring sportscasters hone their skills on their own while watching their favorite sports event by turning down the sound on their televisions and tape-recording their own play-by-play deliveries.

Employers

Most sports broadcasters work for television networks or radio stations. The large sports networks also employ many broadcasters. John Earnhardt says, "The main employers of sports broadcasters are sports networks that own the rights to broadcast sporting events and the broadcast stations themselves." Radio sportscasters are hired by radio stations that range from small stations to mega-stations.

Sports announcers work for professional sports arenas, sports teams, minor league and major league ball teams, colleges, universities, and high schools.

Because sports are popular all over the country, there are opportunities everywhere, although the smaller the town the fewer the opportunities. "Larger cities generally have more opportunities because of the number of stations and the number of sports teams that need to be covered," Earnhardt says.

Starting Out

Although an exceptional audition tape might land a beginner an on-camera or on-the-air job, most sportscasters get their start by writing copy, answering phones, operating cameras or equipment, or assisting the sportscaster with other jobs. Internships or part-time jobs that give the beginner the opportunity to grow comfortable in front of a camera or behind a microphone are invaluable experiences. Of course, contacts within the industry come in handy. In many cases, it is simply an individual's devotion to the sport and the job that makes the difference—that and being in the right place at the right time. John Earnhardt adds that knowledge is key as well. "It obviously helps to know the sport you are reporting on—first, one needs to study the sport and know the sport's rules, history, and participants better than anyone," he advises.

Don't forget to put together an audio tape (if you are applying for a radio job or an announcer position) or a video tape (for television jobs) that showcases your abilities. On the tape, give your real-live account of the sports events that took place on a certain day.

Advancement

In the early stages of their careers, sportscasters might advance from the position of "gofer" to a sports copywriter to actual broadcaster. Later in their careers, sportscasters advance by moving to larger and larger markets, beginning with local television stations and advancing to one of the major networks.

Sportscasters who work in radio may begin in a similar way; advancement for these individuals might come in the form of a better time slot for a sports show, or the chance to give more commentary.

Sports announcers advance by adding to the number of teams for whom they provide public address announcements. Some sports announcers also may start out working for colleges and minor leagues and then move up to major league work.

Earnings

Salaries in sportscasting vary, depending on the medium (radio or television), the market (large or small, commercial or public), and whether the sportscaster is a former athlete or recognized sports celebrity, as opposed to a newcomer trying to carve out a niche.

According to the *Occupational Outlook Handbook*, the average salaries of television sportscasters range from $68,900 for weekday anchors to $37,200 for weekend sportscasters.

Sportscasting jobs in radio tend to pay less than those in television. Beginners will find jobs more easily in smaller stations, but the pay will be correspondingly lower than it is in larger markets. The average salary for a radio sportscaster, according to a 2001 survey by the National Association of Broadcasters and the Broadcast Cable Financial Management Association, was $53,300.

Salaries are usually higher for former athletes and recognized sports personalities or celebrities, such as ex-coaches like John Madden. These individuals already have an established personality within the sports community and may thus have an easier time getting athletes and coaches to talk to them. Salaries for such recognizable personalities can be as high as $2,000,000 a year.

Work Environment

Sportscasters usually work in clean, well-lighted, sound-proof booths or sets in radio or television studios, or in special sound-proof media rooms at the sports facility that hosts sports events. Depending on the sportscaster's scheduled broadcasts, he or she may work off hours, but no matter when the broadcasts are scheduled, it makes for a long day.

Time constraints and deadlines can create havoc and add stress to an already stressful job; often a sportscaster has to race back to the studio to make the final evening broadcast. Sportscasters who deliver play-by-play commentary for radio listeners have the very stressful job of describing everything going on in a game as it happens. They can't take their eyes off the ball and the players while the clock is running and this can be nerve-wracking and stressful.

On the other hand, sportscasters are usually on a first-name basis with some of the most famous people in the world, namely, professional athletes. They quickly lose the star-struck quality that usually afflicts most spectators and must learn to ask well-developed, concise, and sometimes difficult questions of coaches and athletes.

Sports announcers usually sit in press boxes near the action so they can have a clear view of players and their numbers when announcing. Depending on the type of sport, this may be an enclosed area or they may be out in the open air. Sports announcers start announcing before the event begins and close the event with more announcements, but then are able to end their work day. Because sporting events are scheduled at many different times of the day, announcers sometimes must be available at odd hours.

Outlook

Competition for jobs in sportscasting will continue to be fierce, with the better-paying, larger-market jobs going to experienced sportscasters who have proven they can keep ratings high. Sportscasters who can easily substitute for other on-camera newscasters or anchors may be more employable.

The projected outlook is one of slower than average growth, as not that many new radio and television stations are expected to enter the market. Most of the job openings will come as sportscasters leave the market to retire, relocate, or enter other professions. In general, employment in this field is not affected by economic recessions or declines; in the event of cutbacks, the on-camera sports broadcasters and announcers are the last to go.

For More Information

For a list of schools that offer programs and courses in broadcasting, contact:
BROADCAST EDUCATION ASSOCIATION
1771 N Street, NW
Washington, DC 20036-2891
Tel: 202-429-5355
Web: http://www.beaweb.org

To get general information about broadcasting, contact:
NATIONAL ASSOCIATION OF BROADCASTERS
1771 N Street, NW
Washington, DC 20036-2891
Tel: 202-429-5300
Email: careercenter@nab.org
Web: http://www.nab.org

For information on FCC licenses, contact:
FEDERAL COMMUNICATIONS COMMISSION (FCC)
445 12th Street, SW
Washington, DC 20554
Tel: 888-225-5322
Email: fccinfo@fcc.gov
Web: http://www.fcc.gov

For career information and helpful Internet links, contact:
RADIO-TELEVISION NEWS DIRECTORS ASSOCIATION
1600 K Street, NW, Suite 700
Washington, DC 20006-2838
Tel: 202-659-6510
Email: rtnda@rtnda.org
Web: http://www.rtnda.org

Talent Agents and Scouts

Quick Facts

School Subjects
Business
Theater/dance

Personal Skills
Communication/ideas
Leadership/management

Work Environment
Primarily indoors
Primarily one location

Minimum Education Level
Bachelor's degree

Salary Range
$18,000 to $50,000 to $100,000+

Certification or Licensing
None available

Outlook
About as fast as the average

Overview

An agent is a salesperson who sells artistic talent. *Talent agents* act as the representatives for actors, directors, writers, models, and other people who work in film, television, and theater, promoting their talent and managing legal contractual business.

History

The wide variety of careers that exist in the film and television industries today evolved gradually. In the 19th century in England and America, leading actors and actresses developed a system, called the "actor-manager system," in which the actor both performed and handled business and financial arrangements. Over the course of the 20th century, responsibilities diversified. In the first decades of the century, major studios took charge of the actors' professional and financial management.

In the 1950s the major studio monopolies were broken, and control of actors and contracts came up for grabs. Resourceful business-minded people became agents when they realized that there was money to be made by controlling access to the talent behind movie and television productions. They became middlemen between actors (and other creative people) and the production studios, charging commissions for use of their clients.

Currently, commissions range between 10 and 15 percent of the money an actor earns in a production. In more recent years, agents have formed revolutionary deals for their stars, making more money for agencies and actors alike. Powerful agencies such as Creative Artists Agency (CAA), International Creative Management (ICM), and The William Morris Agency are credited with (or, by some, accused of) heralding in the age of the multimillion dollar deal for film stars. This has proved controversial as some top actor fees have inflated to over $20 million per picture; some industry professionals worry that high actor salaries are cutting too deeply into film budgets, while others believe that actors are finally getting the fair share of the profits. Whichever the case, the film industry still thrives, and filmmakers still compete for the highest-priced talent. And the agent, always an active player in the industry, has become even more influential in how films are made.

In the 1960s a number of models became popular celebrity figures, such as Jean Shrimpton, Twiggy, and Varushka, who were noted not only for their modeling work but also for the image and lifestyle they portrayed. In the early days of the fashion industry, models were the products of modeling schools, which also monitored their work schedules. However, as the industry grew and individual models became successful, models often needed, and relied, on someone to manage and organize their careers. Thus, modeling agencies developed to fill this niche. Ford Models, Inc., an agency founded in 1946 by Eileen and Jerry Ford, was one of the first modern agencies devoted to promoting the career of the fashion model. The agency made fashion history by negotiating the first big money contract between model Lauren Hutton and Revlon. Today, Ford Models is an industry leader, employing many talented agents and scouts internationally to represent hundreds of the world's top models.

Sports figures, like movie stars and models, have become internationally recognized figures, renowned not only for their athletic prowess, but also for their charismatic personalities. Like movie stars, athletes began to realize the need to have talented representation—or agents—to protect and promote their interests during contract negotiations. In addition, today's sports agents handle most, if not all, aspects of a professional athlete's career, from commercial endorsements to financial investments to postretirement career offers.

The Job

Talent agents act as representatives for actors, writers, artists, models, athletes, and others who work in performing and visual arts, fashion, sports, and advertising. They look for clients who have potential for success and then work aggressively to promote their clients to film and television directors, casting directors, production companies, advertising companies, sports managers, publishers, catalog companies, photographers, galleries, and other potential employers. Agents work closely with clients to find assignments that will best achieve clients' career goals.

Agents find clients in several ways. Those who work for an agency might be assigned a client by the agency, based on experience or a compatible personality. Some agents also work as talent scouts and actively search for new clients, whom they then bring to an agency. Or the clients themselves might approach agents who have good reputations and request their representation. The methods agents use to find talent are different, depending on each specialty. A sports agent follows high school and college sports to find athletes who have good potential for a career in professional sports. Modeling, acting, and broadcasting agents review portfolios, screen tests, and audiotapes to evaluate potential clients' appearance, voice, personality, experience, ability to take direction, and other factors. A literary agent reads scripts, books, articles, short stories, and poetry submitted by writers. An artist's agent looks at portfolios and original works of art, visits galleries, attends art fairs, and visits student exhibitions. All agents consider a client's potential for a long career—it is important to find people who will grow, develop their skills, and eventually create a continuing demand for their talents.

When an agent agrees to represent a client, they both sign a contract which specifies the extent of representation, the time period, payment, and other legal considerations.

When agents look for jobs for their clients, they do not necessarily try to find as many assignments as possible. Agents try to carefully choose assignments that will further their clients' careers. For example, an agent might represent an actor who wants to work in film, but is having difficulty finding a role. The agent looks for roles in commercials, music videos, or voice-overs that will give the actor some exposure. A model's agent might find shooting assignments for fashion catalogs while searching for a high-profile assignment with a beauty and fashion magazine.

Agents also work closely with the potential employers of their clients. They need to satisfy the requirements of both parties. Agents who represent actors have a network of directors, producers, advertising executives, and photographers that they contact frequently to see if any of their clients can meet their needs. Models' agents are in touch with magazine and catalog publishers, advertising firms, fashion designers, and event planners. Literary agents have contacts in the publishing world, including small and large presses, magazines, and newspapers. Sports agents know the management personnel of professional sports teams. Artists' representatives know gallery owners, art dealers, and art book publishers.

When agents see a possible match between employer and client they speak to both and quickly organize meetings, interviews, or auditions so that employers can meet potential hires and evaluate their work and capabilities. Agents must be persistent and aggressive on behalf of their clients. They spend time on the phone with employers convincing them of their clients' talents and persuading them to hire clients. There may be one or several interviews and the agent may coach clients through this process, to make sure clients understand what the employer is looking for and adapt their presentations accordingly. When a client achieves success and is in great demand, the agent receives calls, scripts, and other types of work requests and passes along only those that are appropriate to the interests and goals of the client.

When an employer agrees to hire a client, the agent helps negotiate a contract that outlines salary, benefits, promotional appearances, and other fees, rights, and obligations. Agents have to look out for the best interests of their clients and at the same time satisfy employers in order to establish continuing, long-lasting relationships.

In addition to promoting individuals, agents may also work to make package deals—for example, combining a writer, director, and a star to make up a package, which they then market to production studios. The agent charges a packaging commission to the studio in addition to the commissions agreed to in each package member's contract. A strong package can be very lucrative for the agency or agencies who represent the talent involved, since the package commission is often a percentage of the total budget of the production.

Agents often develop lifelong working relationships with their clients. They act as business associates, advisers, advocates, mentors, teachers, guardians, and confidantes. Because of the complicated nature of these relationships, they can be volatile, so a successful relationship requires trust and respect on both sides, which can only be earned through experience and time. Agents who represent high-profile talent comprise only a small percentage of agency work. Most agents represent lesser-known or locally known talent.

The largest agencies are located in Los Angeles and New York City, where film, theater, advertising, publishing, fashion, and art-buying industries are centered. There are modeling and theatrical agencies in most large cities, however, and independent agents are established throughout the country.

Requirements

HIGH SCHOOL

You should take courses in business, mathematics, and accounting to prepare for the management aspects of an agent's job. Take English and speech courses to develop good communication skills because an agent must be gifted at negotiation. You also need a good eye for talent, so you need to develop some expertise in film, theater, art, literature, advertising, sports, or whatever field you hope to specialize in.

POSTSECONDARY TRAINING

There are no formal requirements for becoming an agent, but a bachelor's degree is strongly recommended. Advanced degrees in law and business are becoming increasingly prevalent; law and business training are useful because agents are responsible for writing contracts according to legal regulations. However, in some cases, an agent may obtain this training on the job. Agents come from a variety of backgrounds; some of them have worked as actors and then shifted into agent careers because they enjoyed working in the industry. Agents who have degrees from law or business schools have an advantage when it comes to advancing their careers or opening a new agency.

OTHER REQUIREMENTS

It is most important to be willing to work hard and aggressively pursue opportunities for clients. You should be detail-oriented and have a good head for business; contract work requires meticulous attention to detail. You need a great deal of self-motivation and ambition to develop good contacts in industries that may be difficult to break into. You should be comfortable talking with all kinds of people and be able to develop relationships easily. It helps to be a good general conversationalist in addition to being knowledgeable about your field.

Exploring

Learn as much as you can about the industry that interests you. If it's film, read publications agents read, such as *Daily Variety* (http://www.variety.com), *The Hollywood Reporter* (http://www.hollywoodreporter.com), *Premiere* (http://www.premiere.com), and *Entertainment Weekly* (http://www.ew.com). See current movies to get a sense of the established and up-and-coming talents in the film industry. Trace the careers of actors you like, including their early work in independent films, commercials, and stage work.

If sports is your interest, watch games and pay attention to the negotiations for players. Read media reports on the management, coaching, and team-building strategies for professional sports.

To explore the world of fashion and modeling, read *Vogue* (http://www.vogue.com), *W,* and other beauty and glamour magazines. Attend fashion shows. Learn about fashion photography.

If you are interested in art or literature, study historical and current trends. There are numerous art and literary review publications in your library and on newsstands. Look for *Art Business Magazine, New Art Examiner* (http://www.newartexaminer.org), *Artweek* (http://www.artweek.com), and *Communication Arts* (http://www.commarts.com).

If you live in Los Angeles or New York, you may be able to volunteer or intern at an agency to find out more about the career. If you live outside Los Angeles and New York, check your phone book's Yellow Pages, or search the Web, for listings of local agencies. Most major cities have agents who represent local performing artists, actors, and models. If you contact them, they may be willing to offer you some insight into the nature of talent management in general.

Employers

Talent agencies are located all across the United States, handling a variety of talents. Those agencies that represent artists and professionals in the film industry are located primarily in Los Angeles. Some film agencies, such as The William Morris Agency, are located in New York City. An agency may specialize in a particular type of talent, such as minority actors, extras, or TV commercial actors. The top three film agencies—CAA, ICM, and The William Morris Agency—employ approximately 1,500 agents. The Association of Authors' Representatives has a list of member agencies and the vast majority are located in New York City. Pro Sports Group offers a comprehensive sports agent directory that includes more than 1,000 certified MLB, NBA, NFL, and NHL agents. The top modeling agencies, such as Wilhelmina, Ford, and Elite are located in New York City, but there are talent/modeling agencies in all metropolitan areas. Check your Yellow Pages for listings.

Starting Out

The best way to enter this field is to seek an internship with an agency. If you live in or can spend a summer in Los Angeles or New York, you have an advantage in terms of numbers of opportunities. Libraries and bookstores will have resources for locating talent agencies. By searching the Web, you can find many free listings of reputable agents. The Screen Actors Guild (SAG) also maintains a list of franchised agents that is available on its Web site. The Yellow Pages will yield a list of local talent agencies. For those who live in Los Angeles, there are employment agencies that deal specifically with talent agent careers. Compile a list of agencies that may offer internship opportunities. Some internships will be paid, others may provide college course credit, but most importantly, they will provide you with experience and contacts in the industry. An intern who works hard and knows something about the business stands a good chance of securing an entry-level position at an agency. At the top agencies, this will be a position in the mail room, where almost everyone starts. In smaller agencies, it may be an assistant position. Eventually persistence, hard work, and cultivated connections will lead to a job as an agent.

Advancement

Once you have a job as an assistant, you will be allowed to work closely with an agent to learn the ropes. You may be able to read contracts and listen in on phone calls and meetings. You will begin to take on some of your own clients as you gain experience. Agents who wish to advance must work aggressively on behalf of their clients, as well as seek out quality talent to bring into an agency. Those who are successful command more lucrative salaries and may choose to open their own agencies. Some agents find that their work is a good stepping stone toward a different career in the industry.

Earnings

Earnings for agents vary greatly, depending upon the success of the agent and his or her clients. An agency receives 10 to 15 percent of a client's fee for a project. An agent is then paid a commission by the agency, as well as a base salary. Assistants generally make low entry-level salaries of between $18,000 and $20,000 a year. In the first few years, an agent will make between $25,000 and $50,000 a year. However, those working for the top agencies can make much more.

Salaries for fashion model agents depending on the agent's experience, the size and location of the agency, and the models represented. A new agent can expect to earn an annual starting salary of approximately $26,000 working for a Chicago-area modeling agency in 2000. An agent with previous agency experience can earn about $45,000 per year or more. Agents at the top of the industry may make in the hundreds of thousands of dollars. Some agencies choose to pay their agents a commission based on fees generated by model/client bookings. These commissions normally range from 10 to 15 percent of booking totals.

The average yearly salary for a sports agent just starting out ranges from $20,000 to $25,000. As the agent acquires more athletes or the status of his or her clients increases, the agent's salary will increase to $35,000 to $40,000 per year. The high end for the typical agent is approximately $40,000 to $60,000 a year. Agent commissions at top sports management firms run anywhere from 5 to 10 percent of the player's earnings, and up to 25 percent for endorsements the agency negotiates on behalf of the athlete.

Literary agents generally earn between $20,000 and $60,000 annually, with a rare few making hundreds of thousands of dollars a year. Commissions range from 4 to 20 percent of their clients' earnings.

Working for an agency, an experienced agent will receive health and retirement benefits, bonuses, and paid travel and accommodations.

Work Environment

Work in a talent agency can be lively and exciting. It is rewarding to watch a client attain success with your help. The work can seem very glamorous, allowing you to rub elbows with the rich and famous, and make contacts with the most powerful people in entertainment, sports, fashion, or publishing. Most agents, however, represent less-famous actors, artists, models, authors, and athletes.

Agents' work requires a great deal of stamina and determination in the face of setbacks. The work can be extremely stressful, even in small agencies. It often demands long hours, including evenings and weekends. To stay successful, agents at the top of the industry must constantly network. They spend a great deal of time on the telephone, with both clients and others in the industry, and attending industry functions.

Outlook

Employment in the arts and entertainment field is expected to grow rapidly in response to the demand for entertainment from a growing population. However, the numbers of artists and performers also continues to grow, creating fierce competition for all jobs in this industry. This competition will drive the need for more agents and scouts to find talented individuals and place them in the best jobs.

The film industry is enjoying record box office receipts. With markets overseas expanding, even the films that don't do so well domestically can still turn a tidy profit. As a result, agents at all levels in the film industry will continue to thrive. Also, more original cable television programming will lead to more actors and performers seeking representation. The fashion and modeling industries fluctuate slightly with the economy. During recession periods, consumers are likely to spend less, and advertisers plan more modest campaigns. There is stiff competition among the vast numbers of hopeful models for the relatively few available positions and it takes skillful agents to find the best assignments for their clients. Sports has become a huge business in the United States and agents are becoming increasingly important in the negotiation of player contracts. Artists' and authors' agents play an important role in getting their clients' work seen and read. Most book publishers will not even consider a manuscript unless it is submitted through a reputable agent.

For More Information

This organization's Web site lists member agencies, offers a newsletter, and provides links to other literary sites.
ASSOCIATION OF AUTHORS' REPRESENTATIVES
PO Box 237201
Ansonia Station
New York, NY 10003
Web: http://www.aar-online.org

Contact this association for insight into a performing arts career.
NATIONAL ASSOCIATION OF PERFORMING ARTS MANAGERS AND AGENTS
459 Columbus Avenue, Suite 133
New York, NY 10024
Tel: 888-745-8759
Email: info@napama.org
Web: http://www.napama.org

For extensive information on becoming a sports agent, contact:
PRO SPORTS GROUP
PO Box 81
Teaneck, NJ 07666
Tel: 888-912-4747
Web: http://www.prosportsgroup.com

Visit the SAG Web site for information about acting in films, and for a list of talent agencies:
SCREEN ACTORS GUILD (SAG)
5757 Wilshire Boulevard
Los Angeles, CA 90036-3600
Tel: 323-954-1600
Web: http://www.sag.com

Television Directors

Overview

Television directors have ultimate control over the decisions that shape a TV production. The director is an artist who coordinates the elements of a television show and is responsible for its overall style and quality. Television directors work on a variety of productions. For example, they may work on local news programs, cover national sporting events, or tape commercials for businesses. And with the development of "narrowcasting" (broadcasting meant for limited viewing, such as for classrooms, hospitals, or corporations),

Quick Facts

School Subjects
  English
  Theater/dance
Personal Skills
  Communication/ideas
  Leadership/management
Work Environment
  Primarily indoors
  Primarily one location
Minimum Education Level
  Bachelor's degree
Salary Range
  $25,920 to $65,000 to $250,000
Certification or Licensing
  None available
Outlook
  Faster than the average

some directors create programming for very small audiences. Directors are well known for their part in guiding actors and other TV professionals, but they are involved in much more—casting, costuming, cinematography, editing, and sound recording. Every television project, no matter how short or how small the intended audience, requires a director.

History

Drama developed in societies all over the world and dates back thousands of years. Playwrights and actors strove to impress, educate, and influence audiences with their dramatic interpretations of stories. From these ancient beginnings until the early half of the 19th century, actors directed themselves in productions. Until that time it had been common practice for one of the actors involved in a production to be responsible not only for his or her own performance but also for conducting rehearsals and coordinating the tasks involved in putting on a play. Usually the most experienced and respected troupe member would guide the other actors, providing advice on speech, movement, and interaction.

A British actress and opera singer named Madame Vestris (1797-1856) is considered to have been the first professional director. In the 1830s Vestris leased a theater in London and staged productions in which she herself did not perform. She displayed a new, creative approach to directing, making bold decisions about changing the traditional dress code for actors and allowing them to express their own interpretations of their roles. Vestris coordinated rehearsals, advised on lighting and sound effects, and chose nontraditional set decorations; she introduced props, such as actual windows and doors, that were more realistic than the usual painted panoramas.

By the turn of the century, theater directors such as David Belasco (1859-1931) and Konstantin Stanislavsky (1863-1938) had influenced the way in which performances were given, provoking actors and actresses to strive to identify with the characters they revealed so that audiences would be passionately and genuinely affected. By the early 1900s, Stanislavsky's method of directing performers had made an overwhelming mark on drama. His method (now often referred to as "the Method"), as well as his famous criticism, "I do not believe you," continue to influence performers to this day.

By the early 20th century, the film industry had also begun its rapid growth, and directors joined the teams of professionals who contributed to the production of movies. Commercial television, first available in the mid-1940s, provided yet another medium for directors, actors, and other professionals to work with. As a variety of television shows developed—dramas, newscasts, talk shows, for example—and the number of television stations multiplied, new opportunities opened up for television directors. Today, some TV directors work on a freelance basis, completing different projects for various production companies; others work for special interest cable channels, such as ESPN or MTV; still others work for network affiliates across the country. No matter what the project, however, all directors have one thing in common: they direct the talents and skills of a number of professionals, bringing together all the pieces to create a complete program.

The Job

The television director's responsibilities are varied and often depend on the project. For example, the director of a TV movie, documentary, or an episode of a series will have more control over the final production than the director of a news broadcast or a live event. The TV director working on a movie will use a script, go through rehearsals with actors, and shoot and reshoot scenes from many perspectives. To achieve the intended mood of the piece—such as dark and gloomy for a mystery—the director carefully orchestrates the work of lighting and sound technicians, camera operators, and editors. With such a production, the director can take the time to polish the final product that the audience will see.

A director covering a live event, such as a football game, has much less control over the outcome. In this case, the director only gets one shot at broadcasting the game. The director's responsibilities may include working with announcers to make sure their equipment is functioning properly; stationing camera operators so that they are positioned correctly to cover any possible play the teams might make; and being ready to introduce graphics, such as charts with player statistics. The director has little room for error when covering a live event.

Richard Perry, a director/editor for WWAY TV-3 in Wilmington, North Carolina, directs the 5:30 PM newscast. *News directors* may need to combine working from a script and arranged order of stories with the unpredictability of covering a live event. In addition, the news director's job isn't limited to the actual broadcast; he or she must also direct promotional segments, news updates, and some videotaped segments to accompany the live reports later. Although Perry directs the evening news, he reports to work at 5 AM to help prepare for the daily morning show as well as to direct the various updates and promotions to be broadcast throughout the day.

One of the first things Perry does every morning is to make the graphics for the morning news and, as he says, "I prep the 'supers'." Supers are superimposed words that run across the TV screen and provide information such as the names of interview subjects. The newsroom sends a printed list of these supers to Perry. He then types the names and titles into the chryon (the character generator), making sure everything is spelled correctly and in the right order. From 6 to 7 AM, Perry runs the chryon for the morning show. This involves hitting the control for running the words across the screen at the right time. From 7 to 8:30 AM, Perry directs the local weather and news cut-ins for broadcasting during "Good Morning, America."

After his morning work, Perry is usually able to take a break. He returns to the station mid-afternoon and often spends a couple of hours working on commercial production, then he begins pre-production for the evening news. "We make graphics for over-the-shoulder shots," he explains, "and put boards on tape for the news editors to put in their stories." Boards are graphics and lists of information. These boards are videotaped, then edited by the reporter directly into news packages (or self-contained stories on tape, consisting of the reporter's audio and edited video). In the half hour before the evening news, Perry goes over the script, which has information about the order stories will be shown and who is covering them.

During the broadcast, the director typically sits in the control room and wears a headset through which he or she can communicate with the producer and TV crew. The director gives orders to keep the broadcast running smoothly and on time. Large TV stations have both a director and a *technical director*—a member of the technical crew who works directly with the cameras and other equipment and may make adjustments on the control board. But in a smaller station, like Perry's, directors take on many responsibilities. "I sit in front of the switcher," Perry says, "and tell everyone what to do and push all the right buttons at the right time so the show looks smooth." He tells the camera people what to shoot next and calls for tapes to be played (or "rolled") during the broadcast.

Freelance directors may work on live or taped productions. In addition to their responsibilities as directors—coordinating the work of crews, deciding on shots, overseeing editing—they may also need to take on many other elements involved in production. For example, they may need to work on getting funding for a project, hiring writers and assistants, or setting up locations for filming. They may even be involved in publicizing the project and entering it in festivals or competitions. In addition to all these responsibilities, freelance directors must also continually promote themselves and look for new projects to work on. The workflow for a freelance director can be unpredictable. They often take on a variety of projects, covering anything from sporting events to beauty pageants, in order to maintain a steady work schedule.

Requirements

HIGH SCHOOL

The sooner you can get to know a camera and how to set up interesting shots, the better. If your high school offers courses about media or television production, be sure to take those. You should also consider taking photography classes that will teach you about the composition of an image. Take English and journalism classes that will hone your communication skills and give you a feeling for completing assignments on deadline. Computer classes that teach you to work with graphics programs will be beneficial. If you are considering working as a freelancer, take mathematics, business, or accounting classes to help you manage your business. If you're interested in live directing and working with actors and story scripts, take drama classes to gain experience in this area.

POSTSECONDARY TRAINING

Although a college degree isn't necessarily required of a TV director, it does give you an edge in the workplace. Also, many colleges have internship programs and career services that can help you get your foot in the door of the professional world. If you're interested in working for a TV news station, you should apply to the broadcast departments of journalism schools.

If you're interested in directing dramas or sitcoms for network and cable TV, you may want to enroll in a drama school to develop a theater background and experience working with scripts and actors. A number of universities and colleges also offer film studies programs or courses on television broadcast production. Your guidance counselor should be able to help you locate these. Also, do research on your own by checking out school Web sites and reading books such as *The Complete Guide to American Film Schools* and *Cinema and Television Courses* by Ernest Pintoff.

No matter what college program you enroll in, however, one of your top goals should be to gain practical hands-on experience through an internship at a TV station. You will probably not be paid for your work, but you may be able to get course credit. Some schools offer internship opportunities.

Organizations may also be a source of information on internships. The Radio-Television News Directors Association, for example, offers a limited number of scholarships and internships. The Directors Guild of America sponsors several training programs. One of these, the New York Assistant Director Training Program, lasts two years and the participants gain experience shooting on locations primarily in the New York City area. Competition for these programs is extremely fierce. A summer fellowship at the International Radio and Television Society offers an all-expense paid program, which includes career-planning advice and practical experience at a New York-based corporation. They also offer a minority career workshop, which brings college students to New York for orientation in electronic media.

OTHER REQUIREMENTS

Those who want to be television directors should be strong leaders. "You have to be in order to pull together so many people to this one common goal: getting the show on the air cleanly," says Richard Perry. Directors also must be able to concentrate in a hectic environment. Perry notes, "I have an ability to focus on whatever is before me and to block out everything else that is unnecessary." A director also needs self-confidence as well as the ability to work with other people. Personality conflicts sometimes arise between producers and directors or other members of the production team. The director needs to be able to mediate differences and bring people together. As a director, you might need to join a union. Directors working at network stations and for major markets typically have union membership.

Exploring

Join your high school newspaper staff to become familiar with reporting and editing. Volunteer to act as a staff photographer. If your high school has its own TV station, join the production crew. You might be able to videotape school events or work on the school newscast. Also, consider getting involved with the drama club. You may not want to be the star of the school play, but you can be involved in production work and may be able to videotape the play for the drama archives.

Contact a local TV station and ask for a tour of the facilities. Explain that you are interested in working as a director and ask to meet with one during your tour. Or, set up a separate appointment for an informational interview with a director. Go to the interview prepared to ask questions about the work and the director's experience. People are often happy to talk about their work if you show a genuine interest.

Employers

Television directors may work as salaried full-time employees for network or network-affiliated stations, cable stations, businesses, or agencies, or they may work as freelancers. Freelancers are not full-time employees of a particular company. Instead, freelancers work on a project-by-project basis for different employers, for example doing one project for a network and another for an advertising agency.

Starting Out

The position of director is not an entry-level job. You will need to work your way up through the ranks, gaining experience and knowledge along the way. Internships provide the best way to enter this competitive field. The internship gives you hands-on experience and the opportunity to make contacts within the industry. Richard Perry's internship as a production assistant during college led to his permanent position with WWAY-TV. "I gradually worked my way up through prompter, camera, tapes, audio, and finally I was a director." Perry also worked part time for the station for three-and-a-half years before being hired full time. Other starting-out possibilities include working as an assistant for a freelance director or video production company. Be prepared to take any position that will give you hands-on experience with cameras and production, even if it's only on a part-time or temporary basis.

Advancement

Advancement for television directors depends somewhat on their individual goals. One director might consider it an advancement to move from general TV programming to special interest programming. Another might feel that becoming a full-time freelancer is an advancement. Those who work at small stations tend to advance by relocating and working for larger stations. "If I want to move up as a director," Richard Perry says, "I'll have to move up to a larger market, maybe Charlotte or Raleigh." Such a move would mean receiving a larger salary and overseeing a bigger staff.

Earnings

Salaries vary greatly for TV directors and are determined by a number of factors. A director of a newscast at a small TV station will probably be at the low end of the scale, while a director working on a hit series for a network may earn hundreds of thousands of dollars a year. A freelance director working project-to-project may earn a great deal one year and much less the following year.

According to the U.S. Department of Labor, the median yearly income for actors, producers, and directors (including TV directors) was $25,920 in 2000. A 2001 salary survey by the Radio-Television News Directors Association found that news directors had salaries that ranged from $21,000 to $250,000. The median annual salary for television news directors was $65,000. Directors who work full-time for stations or other organizations generally receive benefits such as health insurance and paid vacation and sick days.

Work Environment

A TV station is a busy and exciting place where no two days are exactly alike. If the director is working on a live event, the atmosphere may be stressful and somewhat chaotic as he or she makes snap decisions, calls up the correct graphics, and keeps the show within the time limits. A director working on a taped project that will air at a later date may feel somewhat less "on air" stress; however, this director must also constantly pay attention to numerous production details, staying on budget and resolving problems among the staff. Because the director is responsible for clarifying what everyone's responsibilities for a project are, he or she may need to mediate in a tense situation. "Nine-to-five" definitely does not describe a day in the life of a director; 12-hour days (and more) are not uncommon. Because directors are ultimately responsible for so much, schedules often dictate that they become immersed in their work around the clock, from preproduction to final cut. Nonetheless, those able to make it in the industry find their work to be extremely enjoyable and satisfying.

Outlook

The U.S. Department of Labor predicts that employment growth for actors, producers, and directors will increase at a rate faster than the average through 2010. On a cautionary note, those wanting to become directors should realize that many see the television industry as a glamorous field, and there will be stiff competition for jobs.

More TV programs are produced now than ever before, and this number should continue to grow. New technology will allow cable stations to offer hundreds of additional channels and therefore need more original programming. Also, as more businesses and organizations recognize that TV and video productions can educate the public about their work as well as train employees, they will need directors' services to complete new projects.

Newsrooms provide TV stations with healthy profits every year, and this is not expected to change. Therefore, directors will continue to be in demand to direct newscasts. Directors of traditionally less-recognized forms, such as commercials and music videos, are beginning to receive more attention. In 1997, the Emmy Awards program introduced nominations for the best TV commercials of the year. Also, directors of music videos are now listed along with the performer and record company at the beginning of all videos aired on MTV.

In the future, the number of TV directors who work freelance will likely increase. As productions become more costly and as smaller networks produce original programming, hiring directors on a project-to-project basis is becoming more economical.

For More Information

To learn more about the industry and DGA-sponsored training programs and to read selected articles from DGA Magazine, *contact:*
DIRECTORS GUILD OF AMERICA (DGA)
7920 Sunset Boulevard
Los Angeles, CA 90046
Tel: 310-289-2000
Web: http://www.dga.org

For more information on fellowships, contact:
INTERNATIONAL RADIO AND TELEVISION SOCIETY
420 Lexington Avenue, Suite 1601
New York, NY 10170
Tel: 212-867-6650
Web: http://www.irts.org

This organization for electronic media news professionals has information on internships, scholarships, and the news industry. The Web site has a "bookstore" featuring titles of interest to students and professionals involved in the industry.
RADIO-TELEVISION NEWS DIRECTORS ASSOCIATION
1600 K Street, NW, Suite 700
Washington, DC 20006-2838
Tel: 202-659-6510
Email: rtnda@rtnda.org
Web: http://www.rtnda.org

This Web site contains links to numerous television-related sites and lists colleges and universities worldwide that offer training in television broadcast production.
CINEMEDIA
Web: http://www.cinemedia.org

Television Editors

Overview

Television editors perform an essential role in the television industry. They take an unedited draft of videotape and use specialized equipment to improve the draft until it is ready for viewing. It is the responsibility of the television editor to create the most effective product possible. Television editors may also be employed in the film industry. There are approximately 16,000 film and television editors employed in the United States.

Quick Facts

School Subjects
  Art
  English
Personal Skills
  Artistic
  Communication/ideas
Work Environment
  Primarily indoors
  Primarily one location
Minimum Education Level
  Some postsecondary training
Salary Range
  $18,970 to $34,160 to $900,000+
Certification or Licensing
  None available
Outlook
  Faster than the average

History

The television industry has experienced substantial growth in the last few years in the United States. As more people have access to cable television, that industry has grown too. The effect of this growth is a steady demand for the essential skills that television editors provide. With recent innovations in computer technology, much of the work that these editors perform is accomplished using sophisticated computer programs. All of these factors have enabled many television editors to find steady work as salaried employees of television production companies and as independent contractors who provide their services on a per-job basis.

In the early days of the industry, editing was sometimes done by directors, studio technicians, or others for whom this work was not their specialty. Now every videotape, including the most brief television advertisement, has an editor who is responsible for the continuity and clarity of the project.

The Job

Television editors work closely with producers and directors throughout an entire project. These editors assist in the earliest phase, called preproduction, and during the production phase, when actual filming occurs. Their skills are in the greatest demand during postproduction, the completion of primary filming. During preproduction, in meetings with producers, editors learn about the objectives of the film or video. If the project is a television commercial, for example, the editor must be familiar with the product the commercial will attempt to sell. If the project is a feature-length motion picture, the editor must understand the story line. The producer may explain the larger scope of the project so that the editor knows the best way to approach the work when it is time to edit the film. In consultation with the director, editors may discuss the best way to accurately present the screenplay or script. They may discuss different settings, scenes, or camera angles even before filming or taping begins. With this kind of preparation, film and television editors are ready to practice their craft as soon as the production phase is complete.

Feature-length films, of course, take much more time to edit than television commercials. Therefore, some editors may spend months on one project, while others may be working on several shorter projects simultaneously.

Steve Swersky owns his own editorial company in Santa Monica, California, and he has done editing for commercials, films, and TV. In addition to editing many Jeep commercials and coming-attractions trailers for such movies as *Titanic, Fargo,* and *The Usual Suspects,* Swersky has worked on 12 films. Though commercials can be edited quickly, a film project can possibly take six to nine months to edit. Swersky's work involves taking the film that has been developed in labs and transferring it to videotape for him to watch. He uses "nonlinear" computer editing for his projects, as opposed to traditional "linear" systems involving many video players and screens. "The difference between linear and nonlinear editing," he says, "is like the difference between typing and using a word processor. When you want to change a written page, you have to retype it; with word processing you can just cut and paste." Swersky uses the Lightworks nonlinear editing system. With this system, he converts the film footage to a digital format. The computer has a database that tracks individual frames, and puts all the scenes together in a folder of information. This information is stored on a hard drive and can instantly be brought up on a screen, allowing an editor to access scenes and frames with the click of a mouse.

Editors are usually the final decision-makers when it comes to choosing which segments will stay in as they are, which segments will be cut, or which may need to be redone. Editors look at the quality of the segment, its dramatic value, and its relationship to other segments. They then arrange the segments in an order that creates the most effective finished product. "I assemble the scenes," Swersky says, "choosing what is the best, what conveys the most emotion. I bring the film to life, in a way." He relies on the script and notes from the director, along with his natural sense of how a scene should progress, in putting together the film, commercial, or show. He looks for the best shots, camera angles, line deliveries, and continuity.

Some editors specialize in certain areas of television or film. *Sound editors* work on the soundtracks of television programs or motion pictures. They often keep libraries of sounds that they reuse for various projects. These include natural sounds such as thunder or raindrops, animal noises, motor sounds, or musical interludes. Some sound editors specialize in music and may have training in music theory or performance. Others work with sound effects. They may use unusual objects, machines, or computer-generated noisemakers to create a desired sound for a film or TV show.

Requirements

HIGH SCHOOL

Broadcast journalism and other media and communications courses may provide you with practical experience in video editing. Because film and television editing requires a creative perspective along with technical skills, you should take English, speech, theater, and other courses that will allow you to develop writing skills. Art and photography classes will involve you with visual media. If you're lucky enough to attend a high school that offers film classes, either in film history or in film production, be sure to take those courses. Finally, don't forget to take computer classes. Editing work constantly makes use of new technology, and you should become familiar and comfortable with computers as soon as possible.

POSTSECONDARY TRAINING

Some studios require a bachelor's degree for those seeking positions as television or film editors. Yet actual on-the-job experience is the best guarantee of securing lasting employment. Degrees in liberal arts fields are preferred, but courses in cinematography and audiovisual techniques help editors get started in their work. You may choose to pursue a degree in such subjects as English, journalism, theater, or film. Community and two-year colleges often offer courses in the study of film as literature. Some of these colleges also teach video and film editing. Universities with departments of broadcast journalism offer courses in video and film editing and also may have contacts at local television stations.

Training as a television or film editor takes from four to 10 years. Many editors learn much of their work on the job as an assistant or apprentice at larger studios that offer these positions. During an apprenticeship, the apprentice has the opportunity to see the work of the editor up close. The editor may eventually assign some of his or her minor duties to the apprentice, while the editor makes the larger decisions. After a few years, the apprentice may be promoted to editor or may apply for a position as a television or film editor at other studios.

Training in video and film editing is also available in the military, including the Air Force, Marine Corps, Coast Guard, and Navy. You can also take brief courses in Avid technology. The Avid Film Camp (http://www.filmcamp.com) involves college students in course work, labs, and the editing of actual projects.

OTHER REQUIREMENTS

You should be able to work cooperatively with other creative people when editing a project. You should remain open to suggestions and guidance, while also maintaining your confidence in the presence of other professionals. A successful editor has an understanding of the history of television and film and a feel for the narrative form in general. Computer skills are also important and will help you to learn new technology in the field. You may be required to join a union to do this work, depending on the studio.

"You should have a good visual understanding," Steve Swersky says. "You need to be able to tell a story, and be aware of everything that's going on in a frame."

Exploring

Many high schools have film clubs, and some have cable television stations affiliated with the school district. Often school-run television channels give students the opportunity to actually create and edit short programs. Check out what's available at your school.

One of the best ways to prepare for a career as a television or film editor is to read widely. In reading literature, you will develop your understanding of the different ways in which stories can be presented.

You should be familiar with all different kinds of television and film projects, including documentaries, short films, feature films, TV shows, and commercials. See as many different projects as you can and study them, paying close attention to the decisions the editors made in piecing together the scenes.

Large television stations and film companies occasionally have volunteers or student interns. Most people in the industry start out doing minor tasks helping with production. These production assistants get the opportunity to see all of the professionals at work. By working closely with an editor, a production assistant can learn television or film operations as well as specific editing techniques.

Employers

Some television or film editors work primarily with news programs, documentaries, or special features. They may develop ongoing working relationships with directors or producers who hire them from one project to another. Many editors who have worked for a studio or postproduction company for several years often become independent contractors. They offer their services on a per job basis to producers of commercials and films, negotiating their own fees, and typically have purchased or leased their own editing equipment.

Starting Out

Because of the glamour associated with television and film work, this is a popular field that can be very difficult to break into. With a minimum of a high school diploma or a degree from a two-year college, you can apply for entry-level jobs in many television and film studios, but these jobs won't be editing positions. Most studios will not consider people for television or film editor positions without a bachelor's degree or several years of on-the-job experience.

One way to get on-the-job experience is to complete an apprenticeship in editing. However, in some cases, you won't be eligible for an apprenticeship unless you are a current employee of the studio. Therefore, start out by applying to as many television and film studios as possible and take an entry-level position, even if it's not in the editing department. Once you start work, let people know that you are interested in an editor apprenticeship so that you'll be considered the next time one becomes available.

Those who have completed bachelor's or master's degrees have typically gained hands-on experience through school projects. Another benefit of going to school is that contacts that you make while in school, both through your school's placement office and alumni, can be a valuable resource when you look for your first job. Your school's placement office may also have listings of job openings. Some studio work is union regulated. Therefore you may also want to contact union locals to find out about job requirements and openings.

Advancement

Once television and film editors have secured employment in their field, their advancement comes with further experience and greater recognition. Some editors develop good working relationships with directors or producers. These editors may be willing to leave the security of a studio job for the possibility of working one-on-one with the director or producer on a project. These opportunities often provide editors with the autonomy they may not get in their regular jobs. Some are willing to take a pay cut to work on a project they feel is important.

Some editors choose to stay at their studios and advance through seniority to editing positions with higher salaries. They may be able to negotiate better benefits packages or to choose the projects they will work on. They may also choose which directors they wish to work with. In larger studios, they may train and supervise staffs of less experienced or apprentice editors.

Some sound or sound-effects editors may wish to broaden their skills by working as general film editors. Some film editors may, on the other hand, choose to specialize in sound effects, music, or some other editorial area. Some editors who work in television may move to motion pictures or may move from working on commercials or television series to television movies.

Earnings

Television and film editors are not as highly paid as others working in their industry. They have less clout than directors or producers, but they have more authority in the production of a project than many other film technicians. According to the U.S. Department of Labor, the median annual wage for television and film editors was $34,160 in 2000. A small percentage of television and film editors earn less than $18,970 a year, while some earn over $71,280. The most experienced and sought after television and film editors can command much higher salaries. The Avid Film Camp advises students that they may make as little as $15,000 a year in the career; but the camp notes that there are some editors at the top of the field who make over $900,000 a year.

Work Environment

Most of the work done by editors is done in television and film studios or at postproduction companies using editing equipment. The working environment is often a small, cramped studio office. Working hours vary widely depending on the project. During the filming of a commercial, for instance, editors may be required to work overtime, at night, or on weekends to finish the project by an assigned date. Many feature-length films are kept on tight production schedules that allow for steady work unless filming gets behind.

"As stressful as the work can be," Steve Swersky says, "we joke around that it's not like having a real job. Every day is a fun day."

During filming, editors may be asked to be on hand at the filming location. Locations may be outdoors or in other cities, and travel is occasionally required. More often, however, the television or film editor edits in the studio, and that is where the bulk of the editor's time is spent.

Disadvantages of the job involve the editor's low rank on the totem pole of film or television industry jobs. However, most editors feel that this is outweighed by the advantages. Television and film editors can view the projects on which they have worked and be proud of their role in creating them.

Outlook

The outlook for television and film editors is very good. In fact, the U.S. Department of Labor predicts faster than average employment growth for television and film editors through 2010. The growth of cable television and an increase in the number of independent film studios will translate into greater demand for editors. This will also force the largest studios to offer more competitive salaries in order to attract the best television and film editors.

The digital revolution will greatly affect the editing process. Already, there are 20,000 Avid media systems worldwide. Editors will work much more closely with special effects houses in putting together projects. When using more effects, television and film editors will have to edit scenes with an eye towards special effects to be added later. Digital editing systems are also available for home computers. Users can feed their own digital video into their computers, then edit the material, and add their own special effects and titles. This technology may allow some prospective editors more direct routes into the industry, but the majority of editors will have to follow traditional routes, obtaining years of hands-on experience.

For More Information

The ACE features some career and education information for film and television editors on its Web page, along with information about internship opportunities and sample articles from Cinemeditor Magazine.

AMERICAN CINEMA EDITORS (ACE)
100 Universal City Plaza
Building 2282, Room 234
Universal City, CA 91608
Tel: 818-777-2900
Email: amercinema@earthlink.net
Web: http://www.ace-filmeditors.org

This union counts film and television production workers among its craft members. For education and training information as well as links to film commissions and production companies, check out the IATSE Web site's Craft page: Film and Television Production.

INTERNATIONAL ALLIANCE OF THEATRICAL STAGE EMPLOYEES, MOVING PICTURE TECHNICIANS, ARTISTS AND ALLIED CRAFTS OF THE UNITED STATES AND CANADA (IATSE)
1430 Broadway, 20th Floor
New York, NY 10018
Tel: 212-730-1770
Web: http://www.iatse.lm.com

For information on NATAS scholarships and to read articles from Television Quarterly, *the organization's official journal, visit the NATAS Web site.*

NATIONAL ACADEMY OF TELEVISION ARTS AND SCIENCES (NATAS)
111 West 57th Street, Suite 1020
New York, NY 10019
Tel: 212-586-8424
Web: http://www.emmyonline.org

Television Producers

Overview

Television producers are behind-the-scenes professionals who may be involved with budgeting and financing, working out a production timeline, casting appropriate actors, or even hiring the crew for a TV project. Producers oversee the production of newscasts, sporting events, dramas, comedies, documentaries, holiday specials, and the many other programs that make up network and cable broadcasting. Because of the varied nature of television programming, a producer's role may also change from project to project. A producer on one project, for example, may only be involved in arranging financial backing and putting together the creative team of directors and actors. A producer on another project may oversee practically every detail of the production, including arranging for equipment and scheduling personnel.

History

The position of television producer has its roots in two older forms of visual arts—drama and film. Drama has existed in cultures around the world for thousands of years. As drama and theaters developed, the roles of actors, writers, directors, and producers became clearly defined. The motion picture industry, which experienced rapid growth during the early 20th century, provided a new medium for artistic professionals to work in, and many positions, such as producer, were copied from the theater world.

In 1945, following World War II, commercial television broadcasting became available in the United States. The television industry experienced phenomenal growth through increasingly better equipment, more TV stations, and larger audiences. By 1949, for example, the inauguration of President Truman was seen on television by some 10 million people.

Naturally, professionals were needed to create, perform, and produce programming for this ever-growing new media. This new opportunity drew producers from the film world and others who simply wanted to work with television. The producer became an essential member of production teams at small stations across the country as well as at large network stations. Today, with such factors as increasingly sophisticated equipment (for example, computers that can generate images), enormous salaries for popular TV stars (Helen Hunt, for example, made $22 million for her last season of *Mad About You*), and a growing number of networks showing specialized programming (such as Oxygen and VH1), the job of the producer has become more complex. The producer must have a depth of technical knowledge, the ability to manage large financial sums, and an instinct for choosing projects that will draw large audiences.

The Job

Producers oversee projects, from the idea stage to the final taped or aired version. Their responsibilities, however, vary depending on both the producer's employer and the project. One of the primary responsibilities of *independent producers,* those with their own production companies, may be to raise money for projects. A producer at a television station, on the other hand, may be given a budget to work with. In either case, though, the producer is responsible for keeping an eye on costs and making sure the project stays within budget. This also requires time-management skills, because the producer must schedule just the right amount of time for different phases of production. If the producer underestimates the time needed for filming on location, for example, the producer's company will need to pay the expenses for extra time on location, causing the project to go over budget.

A producer's responsibilities are also affected by what type of project he or she works on. For example, *newscast producers,* along with reporters, determine what stories are worth broadcasting. Newscast producers assign stories, review taped reports, and may even help edit the material. Often these producers must deal with late-breaking developments and must quickly assign reporters and TV crews to cover a story, then weave the new report into the broadcast while staying within the broadcast's time requirements. The newscast is a combination of live and taped segments, and the producer often needs to make decisions quickly while the show is on the air.

Documentary producers are also very actively involved in their productions, but they typically have days, rather than hours, to complete projects. They may be involved in deciding on a concept for the documentary; hiring writers, directors, and the crew; and scouting out locations and finding interview subjects. Once interviews and other segments are taped, they may review the material, select the best footage, and edit it into a program of predetermined length.

Whereas newscast and documentary producers rely on their news judgment, *drama and comedy producers* rely on their understanding of the entertainment world. These producers come up with ideas for shows; hire writers, directors, and actors; and review the final product for its tone and content.

Freddy James is an associate producer for Home and Garden Television (HGTV). He works on producing specials for this cable network, and his work often takes him away from the station for periods of time. One project he worked on focused on the Habitat for Humanity program, which recruits volunteers to build homes for families in need. James's work began long before the filming took place. "We pre-interviewed 10 families who were to receive homes and selected three [to focus on]," he says. "We centered the show around them and around the volunteers who came from all over the world to help them build these homes in one week."

With other members of the production, James planned the shoot. They set up all the hotel and crew arrangements, as well as researched the Habitat for Humanity program. They scheduled interviews with officials of Habitat for Humanity. "And we went to the families' old homes and interviewed them about the project and how they felt about being a part of it," James explains. During the four-day shoot, the production team got footage of the houses going up, and they talked to officials and volunteers. The on-air talent then arrived for a day to shoot his commentary.

Once all the filming was done, the producers returned to the station with all the footage and reviewed the material. "We select our bites," James says, "and start writing the show. Our executive producers proof our scripts, then we set up a voice-over session for our talent, and he voices the script." After selecting video for a program, the producers have it digitized for editing on the computer. On this project, the producers also worked with musicians who composed music for the show and graphic artists who created design work.

No matter what type of programming producers work on, they must be organized team leaders, able to communicate their ideas about a project from concept to conclusion.

Requirements

HIGH SCHOOL

Does working as a producer sound interesting to you? If so, there are a number of classes you can take in high school to help you get ready for this work. It will be very important for you to take composition, speech, and English classes to help you develop your research, writing, and speaking skills. These communication skills will be essential to have in your career. In addition, take mathematics, business, or accounting classes since you will be managing budgets. Consider taking psychology classes that may give you an understanding of people and their interests. Take computer classes so that you can develop a familiarity with this tool. If your high school offers any classes on the history of broadcast media or the use of broadcast media, be sure to take those. If you are specifically interested in producing entertainment shows, consider taking drama classes that will give you an understanding of scripts and working with actors.

POSTSECONDARY TRAINING

While there are no formal educational requirements for becoming a television producer, many producers do have college degrees. The degree you get may depend on the area of television that you are interested in. For example, if you are interested in news broadcasting, you may want to consider attending a journalism school to receive a broadcast journalism degree. Make sure that you get a broad-based education, however. Anyone working in the news industry will need an understanding of history, geography, and political science, so take classes that cover these subjects as well. If you are interested in producing series television, you may want to attend a drama or film school. Programs at these schools will help you hone your understanding of story lines and audiences and markets. A number of universities and colleges also offer film studies programs or courses on television broadcast production. Ask your school guidance counselor to help you locate these programs. In addition, do your own research. Check out the Web sites of schools you are interested in and read books on the topic, such as *The Complete Guide to American Film Schools* and *Cinema and Television Courses* by Ernest Pintoff. A number of producers have rounded out their education by taking business courses or even by completing a business degree along with their broadcast studies to prepare them for this career.

Perhaps the most important thing to consider when selecting a school is the internship opportunities that will be available to you. Some schools have their own broadcasting stations where you can work; others require students to complete an internship that they locate on their own; and still others may offer internships through local participating network affiliates. No matter what the arrangements, you should be sure the school you choose will provide you with a way to gain practical hands-on experience working at a TV station. Competition for internships is high, and most positions are unpaid. In some cases, though, you may be able to get course credit for your work.

A number of organizations provide information on internships or sponsor internship programs. The Radio-Television News Directors Association, for example, offers several scholarships and internships for college students involved with electronic journalism. The Institute on Political Journalism offers a seven-week summer program during which college students combine classroom work at Georgetown University with work at a Washington, DC, media organization. The Directors Guild of America sponsors a Los Angeles-based training program for a limited number of college graduates. The trainees in this program are paid and work on television series projects.

OTHER REQUIREMENTS

The successful producer is an organized individual who can deal quickly and effectively with problems that may cause a change in production plans. Producers need to have a good sense of what stories, news, or other items will interest viewers. They also need salesmanship qualities since they may have to "sell" a station on a project idea or convince an actor to take a role. Producers work with teams of professionals and must be able to bring people together to work on the single goal of completing a project. While working in the television world may seem glamorous, you should realize that as a producer your work will take place behind the scenes. Although people recognize stars of sitcoms and news anchors when they are in public, few are able to spot a producer in the crowd. You must be self confident and be comfortable with this anonymity if you are to enjoy this work.

Exploring

Consider joining your high school newspaper's staff. You will get experience at completing projects on a deadline, and, if you are involved in selling ad space, you will get a taste of what it is like to work with budgets and financing. If your school has a radio or TV station, volunteer to be part of the staff. Become involved in the production work and you may learn how to use the cameras, control sound, or edit pieces. If your school doesn't have these facilities, consider joining the drama club. You will be able to do production work for a play and may also gain experience with advertising and financing the production.

To meet professionals in the field, ask your media department teacher or guidance counselor to arrange for a producer from a local TV station to come talk to interested students. Another option is to call the local TV station and request a tour of the facility. Explain that you are interested in becoming a producer and ask to meet a producer during your tour. You may also be able to set up an informational interview with the producer. Come to this interview prepared to ask questions. What type of educational background does this producer have? What is the hardest part of the job? What is the most enjoyable? Does the producer have any advice for you? People are often happy to talk about their work when you have specific questions to ask and show a true interest.

You may even find that you can develop a mentor relationship through this contact.

Employers

Producers may be salaried employees of television stations or networks, or they may work independently, running their own production company. Independent producers may take on projects for networks as well as other organizations. Producers who work on newscasts are employed across the country at large as well as medium-sized and small stations. Producers who work on television series typically live in Los Angeles, where there are network headquarters; some live in New York.

Starting Out

The job of producer is not an entry-level position. You will need to "pay your dues" by gaining experience working with different equipment and people. Internships are often the best way to break into the business. Interns are frequently offered paid positions after they graduate from school. If your internship doesn't lead to employment, make yourself visible in the field and take any job that allows you to work as a member of the production team. Freddy James got his first job his junior year of college after a tour of a TV station. "I made sure the executive producer knew who I was when I left that day," he says. While most of the other students were interested in anchoring, James stressed his interest in producing. "And I sent a follow-up letter to the executive producer. The next thing I knew, they called wanting to know if I wanted to run a studio camera for minimum wage." The station was number one in the city, and James took the job knowing it would be an excellent opportunity to get a start in the field.

Advancement

Advancement for producers depends to an extent on their individual goals. They may change stations, moving from smaller stations to larger ones where they oversee a larger staff and have increased pay. Some producers may decide to specialize, working in an area that interests them such as live TV or music videos. Most consider it an advancement to work on their own projects. At some stations, an *executive producer* will oversee the work of one or more producers. Independent producers may advance by increasing their clientele and working on large-budget projects.

Earnings

Although some producers with their own production companies and some producers working for networks can make well over $100,000 a year, most producers make considerably less. Bringing in the big paychecks as a producer for television comes from years of hard work, good luck, and well-established connections. But good producers don't generally enter the field with big money in mind—they love the work. Producers who don't make a great deal of money benefit in other ways: they often call all the shots and have control of a project.

Salaries for producers depend on such factors as the station's size and location, the type of programming the producer works on, and the producer's experience. The U.S. Department of Labor reported that the median annual salary for actors, directors, and producers (including TV producers) was $25,920 in 2000. According to a 2001 salary survey by the Radio-Television News Directors Association, news producers at the high end of the salary range had yearly earnings of $75,000, while those in the middle averaged $29,300.

Producers who are full-time, salaried employees of stations or networks typically receive benefits such as health insurance and paid vacation and sick days. Independent producers must provide these extras for themselves.

Work Environment

TV stations are generally clean, comfortable places equipped with up-to-date, sophisticated broadcasting equipment. The atmosphere is busy and exciting, with numerous professionals concentrating on completing their specific jobs. Some producers, such as Freddy James, may need to travel and work on location. The conditions for those working on location vary. For example, the producer may have to stand in the rain for hours with the crew attempting to get the right shot. The work of a producer is stressful, both for those who work on live programming and those who work on taped programming. The producer needs to be a creative and fast-acting problem solver because production difficulties arise almost every day. A lot of people are needed to put on a television show, and personalities sometimes clash, which can lead to tension on the set. The producer must be able to work successfully in this atmosphere. A producer's schedule will depend on the projects he or she is working on. A producer of a morning news show, for example, may need to be at the station by 5 AM. A producer working on a television movie, on the other hand, might not need to be at the shoot until midday but work until late in the night. No matter what the project, though, long workdays (10 to 12 hours) are common.

Outlook

According to the U.S. Department of Labor, the employment of actors, directors, and producers is expected to increase faster than the average through 2010. New cable networks needing original programming are developing at a rapid rate. Freelance opportunities are also expected to increase as networks look to independent production companies for more programming. Producers will also be in demand to put together newscasts: newsrooms provide TV stations with healthy profits every year. Because many consider this exciting and glamorous work, however, competition for jobs will be high. TV producers (and those just starting out in the business) who have a thorough knowledge of new technologies will be in the best position to get jobs. Producers will not only need to understand various computer-assisted techniques, but, as broadcasting becomes more closely involved with the Internet and interactive television, producers will also need to know how to work with these technologies.

For More Information

For more information on the industry and training programs, contact:
DIRECTORS GUILD OF AMERICA
7920 Sunset Boulevard
Los Angeles, CA 90046
Web: http://www.dga.org

For more industry news and career information, contact:
PRODUCERS GUILD OF AMERICA
6363 Sunset Boulevard, 9th Floor
Los Angeles, CA 90028
Tel: 323-960-2590
Web: http://www.producersguildonline.com

*RTNDA, an organization for electronic media news professionals,
has information on internships, scholarships, and the news industry.*
**RADIO-TELEVISION NEWS DIRECTORS ASSOCIATION
(RTNDA)**
1600 K Street, NW, Suite 700
Washington, DC 20006-2838
Tel: 202-659-6510
Email: rtnda@rtnda.org
Web: http://www.rtnda.org

*For more information on the Institute on Political Journalism at
Georgetown University, contact:*
THE FUND FOR AMERICAN STUDIES
1706 New Hampshire Avenue, NW
Washington, DC 20009
Tel: 800-741-6964
Web: www.tfas.org

To read television news online, check out this Web site:
THE HOLLYWOOD REPORTER.COM
Web: http://www.hollywoodreporter.com

Weather Forecasters

Overview

Weather forecasters compile and analyze weather information and prepare reports for daily and nightly TV or radio newscasts. Forecasters, also known as *meteorologists* and *weathercasters,* create graphics, write scripts, and explain weather maps to audiences. They also provide special reports during extreme weather conditions. To predict future weather patterns and to develop increased accuracy in weather study and forecasting, forecasters may conduct research on such subjects as atmospheric electricity, clouds, precipitation, hurricanes, and data collected from weather satellites. Other areas of research used to forecast weather may include ocean currents and temperature.

History

Meteorology—the science that deals with the atmosphere and weather—is an observational science, involving the study of such factors as air pressure, climate, and wind velocity. The basic weather instruments had all been invented hundreds of years ago. Galileo (1564-1642) invented the thermometer in 1593, and Evangelista Torricelli (1608-47) invented the barometer in 1643. Simultaneous comparison and study of weather in different areas was impossible until the telegraph was invented. Observations of the upper atmosphere from balloons and airplanes started after

World War I. Not until World War II, however, was great financial support given to the development of meteorology. During this war, a very clear-cut relationship was established between the effectiveness of new weapons and the atmosphere.

More accurate instruments for measuring and observing weather conditions, new systems of communication, and the development of satellites, radar, and high-speed computers to process and analyze weather data have helped weather forecasters and the general public to get a better understanding of the atmosphere.

The Job

El Niño. F5-rated tornadoes storming down tornado alley. Heat waves and ice storms. Flood-started fires in North Dakota. Hurricanes Andrew, Hugo, and Betsy. Extreme weather conditions often become national celebrities while the citizens of the threatened cities suffer. These people look to TV and radio weather forecasters to advise them of upcoming storms, how to prepare for them, and how to recover from them. But weather forecasters aren't just on the air during extreme conditions—they're on radio and TV broadcasts many times every day. On one day people may be relying on their local forecaster to help them prepare for a midnight tornado, on another day they may simply want to know whether to leave the house with an umbrella.

Some weather forecasters are reporters with broadcasting degrees, but over half of TV and radio weather forecasters have degrees in meteorology. Colleges across the country offer courses and degrees in meteorology for people who want to work for broadcast stations, weather services, research centers, flight centers, universities, and other places that study and record the weather. With a good background in the atmospheric sciences, broadcast weather forecasters can make informed predictions about the weather and can clearly explain these predictions to the public. The people of Myrtle Beach, South Carolina, look to Ed Piotrowski of WPDE-TV for the daily weather news as well as information on extreme weather conditions such as approaching hurricanes. As chief meteorologist for WPDE, Piotrowski is responsible for delivering the forecast for radio and four television evening newscasts. "I also manage a staff and several interns," he notes.

Preparing the forecast means interpreting a great deal of data from a variety of different sources. The data may come from various weather stations around the world. Even the weather conditions swirling over the oceans can affect the weather of states far inland, so local weather forecasters keep track of the weather affecting distant cities. In addition, weather stations and ships at sea record atmospheric measurements, information that is then transmitted to other weather stations for analysis. This information makes its way to the National Weather Service in Washington, DC, where scientists develop predictions to send to regional centers across the country. The tools used by meteorologists include weather balloons, instrumented aircraft, radar, satellites, and computers. Instrumented aircraft are high-performance airplanes used to observe many kinds of weather. Radar is used to detect rain or snow as well as other weather. Doppler radar can measure wind speed and direction. It has become the best tool for predicting severe weather. Satellites use advanced remote sensing to measure temperature, wind, and other characteristics of the atmosphere at many levels. Scientists can observe the entire surface of the earth with satellites. The introduction of computers has forever changed the research and forecasting of weather. The fastest computers are used in atmospheric research and for large-scale weather forecasting. Computers are used to produce simulations of upcoming weather.

At Piotrowski's station, the weather center receives information from the National Center for Environmental Prediction, the Storm Prediction Center, and the Severe Storms Forecast Center as well as the National Weather Service. He explains, "All the data collected is put into many computer programs with various scientific formulas. These programs eventually put out weather scenarios for several different times in the future. Our [the forecaster's] job is to interpret these and make our forecast accordingly."

Broadcast weather forecasters may also prepare maps and graphics to aid the viewers. Broadcasting the information means reading and explaining the weather forecast to viewers and listeners. The broadcast weather forecaster must be able to concentrate on several different tasks at once. For example, when Piotrowski broadcasts to the TV audience he is actually standing in front of a plain blue wall (called a "chromakey") that shows graphics only to the viewers. In order for Piotrowski to see the map, he must watch the station's monitor. He then points to areas on the blue wall based on what he sees in the monitor; to the audience, it looks as though he is pointing to places on the map. While talking about the forecast, Piotrowski may have to listen to time cues (the amount of time left for the presentation) from the newscast producer through a hearing device placed in his ear. Throughout all this activity, the broadcast weather forecaster must stay focused and calm.

Many people look to TV and radio news for weather information to help them plan events and vacations. Farmers are often able to protect their crops by following weather forecasts and advisories. The weather forecast is a staple element of most TV and radio newscasts. Some cable and radio stations broadcast weather reports 24 hours a day; most local network affiliates broadcast reports during morning, noon, and evening newscasts, as well as provide extended weather coverage during storms and other extreme conditions. Piotrowski realizes that his responsibilities are vitally important during hurricanes. He explains, "We need to interpret the data to give people the scenario we think will play out in our area upon landfall. It could be the difference between life and death."

In addition to broadcasting weather reports, radio and TV weather forecasters often visit schools and community centers to speak on weather safety. They are also frequently involved in broadcast station promotions, taking part in community events.

Requirements

HIGH SCHOOL
While you are in high school, you can prepare for your future work as a broadcast weather forecaster by taking a number of different classes. Concentrate on the sciences—earth science, biology, chemistry, and physics—to give you an understanding of the environment and how different elements interact. Geography and mathematics courses will also be useful to you. To familiarize yourself with computers and gain experience working with graphics programs, take computer classes. Computers will be an essential tool that you'll use throughout your career. Take plenty of English and speech classes. As a broadcaster you will need to have excel-

lent writing and speaking skills. If your school offers any media courses in which you learn how to broadcast a radio or television show, be sure to take these classes.

POSTSECONDARY TRAINING

Although a degree in meteorology or atmospheric science isn't required to enter the profession, it is necessary for advancement. The American Meteorological Society (AMS) publishes a listing of schools offering degree programs in atmospheric and related sciences. Check your local library for a copy or purchase one from the AMS (see contact information at end of article). These programs typically include such courses as atmospheric measurements, thermodynamics and chemistry, radar, cloud dynamics, and physical climatology. While in college you should also continue to take English or communication classes to hone your communication skills and computer classes to keep up to date with this technology. It is also important, during your college years, to complete an internship as a student weather forecaster with a TV or radio station. Your college and organizations such as the AMS can help you locate an internship. Although it may not pay much—if anything—this position will allow you to have contact with professionals and give you hands-on experience.

CERTIFICATION OR LICENSING

The AMS and the National Weather Association (NWA) each offer certification to broadcast meteorologists. To qualify for the AMS Seal of Approval, applicants must meet educational requirements and demonstrate scientific competence and effective communication skills. Applicants must submit taped samples of their work for review by an evaluation board. Requirements for the NWA Seal of Approval are similar, but the applicant must also have a certain amount of on-air experience and pass a written exam. Although the AMS and the NWA seals are not required for broadcast weather forecasters to work in this business, the holder of these seals will have an advantage when looking for a job. Some TV and radio stations note that their forecaster has a seal of approval in their advertisements.

OTHER REQUIREMENTS

Broadcast weather forecasters must be able to work well under pressure in order to meet deadlines for programming or plot severe weather systems. They must be able to communicate complex theories and events in a manner that is easy for the audience to understand. And, naturally, weather forecasters must have an interest in weather and the environment. "Don't go into TV weather just to be on TV," Ed Piotrowski advises. "You must have a passion for the weather to do a fantastic job."

Exploring

There are several ways you can explore different aspects of this career while you are still in school. Consider joining a science club that is involved in environmental activities. Also, volunteer to work at your school's radio or TV station. You'll learn the basics of putting on a program and might even get on the air yourself. If your school doesn't have one of these stations, join the newspaper staff to get some experience working with the media.

Each year, the National Weather Service (NWS) accepts a limited number of student volunteers, mostly college students but also a few high school students. Local offices of the NWS also allow the public to come in for tours by appointment. Your guidance counselor may be able to help you with this.

Also, contact your local radio and TV stations and ask for a tour. Tell them you are interested in broadcast meteorology and ask to meet the weather forecasting staff. You may be able to arrange for an informational interview with the weather forecaster, during which you might ask that person about his or her education, experiences, the best part of the work, and any other questions you may have.

Employers

Most broadcast weather forecasters work for network television affiliates and local radio stations. Because evening national newscasts do not have weather forecasts, there are fewer network opportunities for broadcast meteorologists. National cable networks, such as The Weather Channel and 24-hour news channels, hire weather forecasters and may offer internships.

Those who have degrees in meteorology or atmospheric science can work for a variety of other services as well—the United

States government is the largest employer of meteorologists in the country. Meteorologists work for the NWS, the military, the Department of Agriculture, and other agencies.

Starting Out

Your college's placement office is a good resource to use when you are looking for your first job. Many people also find their first positions either through connections they have made while interning or through the internship itself. Ed Piotrowski's first full-time job was with the TV station where he interned. "I was very fortunate," he says. "I showed great initiative and the weekend job opened up right around the time I graduated."

Local ads and job listings on the Internet are other sources you should check out. The AMS and the NWA send members job listings and also post openings on their Web sites.

Advancement

Someone forecasting for a network affiliate in a smaller region may want to move to a larger city and a larger audience. "Many weather people strive to get to the big cities to make the big money," Ed Piotrowski says, "but getting to the big markets can be hard and difficult to stay in."

In many cases, meteorologists work up within one station. Full-time broadcast meteorologists generally start forecasting for the weekend news or the morning news, then move up to the evening news. A meteorologist may then become chief meteorologist, in charge of a newscast's weather center and staff. With each advancement comes more responsibility and a larger salary.

Earnings

In the newsroom, weather forecasters generally make more than sportscasters but less than lead anchors. The salary for weather forecasters varies greatly according to experience, region, and media. Those working in television typically earn more than those working in radio. According to a 2001 salary survey by the Radio-Television News Directors Association, weathercasters earned salaries that ranged from $15,000 to $390,000, with a median of $44,500. Radio weather forecasters in small markets may make less than that amount.

According to the U.S. Department of Labor, the median income for meteorologists in nonsupervisory, supervisory, and managerial positions employed by the federal government was $68,100 in 2001.

At a few top TV stations in large cities, a weather forecaster can become something of a local celebrity, attracting higher ratings for the station. Popular weather forecasters with large audiences have been known to command over $400,000 a year, but such salaries are rare.

Work Environment

The atmosphere at a radio or TV station can be exciting, fun, and sometimes tense—especially during times of emergency. Forecasters' schedules depend on the times they are scheduled to be on the air. Those working morning shows, for example, may have to be at the station by 4 AM to prepare for the broadcast. The weather forecaster may also make public appearances, giving talks to schools or clubs. This gives the forecaster the opportunity to meet a lot of people and attend events; however, it also makes for busy and varied days. In addition, they have to be prepared to work odd or long hours during times of weather emergencies.

Weather forecasters work with a great deal of specialized equipment, such as computers and radar, as well as the equipment of the broadcast trade, such as the chromakey, hearing devices, and microphones. The weather forecaster is part of a team (along with such professionals as newscasters, producers, and technicians) who work to put the broadcast on the air.

Outlook

The U.S. Department of Labor predicts employment for all atmospheric scientists to increase about as fast as the average through 2010. Usually, meteorologists are able to find work in the field upon graduation, though they may have to be flexible about the area of meteorology and region of the country in which they work.

Positions for broadcast meteorologists, as with any positions in broadcast news, are in high demand. The number of news departments and news staff is expected to increase at a steady rate, but the growing number of graduates looking for work in news departments will keep the field very competitive.

Currently about half of TV and radio weather forecasters do not hold meteorology degrees; with increased competition for work, forecasters without extensive backgrounds in the atmospheric sciences may find it difficult to get jobs. New tools and computer programs for the compilation and analyses of data are constantly being developed by research scientists. To find good positions, future broadcast meteorologists will need a lot of technical expertise in addition to their understanding of weather.

A national fascination with weather may lead to more outlets for broadcast meteorologists. Look for more cable weather information channels like The Weather Channel to develop. Weather disasters are requiring more coverage by news departments. In addition to forecasting, broadcast meteorologists will be involved in reporting about the after-effects of storms and other extreme conditions. Many people look to the Internet for global and regional weather information, so look for broadcast and Internet weather resources to merge. Broadcast meteorologists are becoming more actively involved in developing and maintaining pages on the World Wide Web.

For More Information

This organization has information on meteorology careers, student membership, and education. For more information, contact:
AMERICAN METEOROLOGICAL SOCIETY
45 Beacon Street
Boston, MA 02108-3693
Tel: 617-227-2425
Email: amsinfo@ametsoc.org
Web: http://www.ametsoc.org/AMS

For weather and employment information and links to other weather-related sites, check out the following Web site:

NATIONAL OCEANIC AND ATMOSPHERIC ADMINISTRATION
14th Street and Constitution Avenue, NW, Room 6013
Washington, DC 20230
Tel: 202-482-6090
Web: http://www.noaa.gov

To read selected sections of the monthly NWA Newsletter *or find out about local chapters, check out the following Web site:*

NATIONAL WEATHER ASSOCIATION (NWA)
1697 Capri Way
Charlottesville, VA 22911-3534
Tel: 434-296-9966
Web: http://www.nwas.org

To learn more about the weather, take a look at the NWS Web site:

NATIONAL WEATHER SERVICE (NWS)
1325 East-West Highway
Silver Spring, MD 20910
Tel: 301-713-0258
Web: http://www.nws.noaa.gov

Index